Improving F ... aracteristics of

The research described in this thesis was performed in the section Physics of Nuclear Reactors (PNR), of the department Radiation, Radionuclides & Reactors (R^3), of the Delft University of Technology, Delft, The Netherlands.
Visiting address: Mekelweg 15, 2629 JB Delft, The Netherlands.

Support: Parts of the work presented in this thesis were financed under the European Commission / EURATOM 6th Framework Programme 'Gas Cooled Fast Reactor Specific Targeted REsearch Programme' (GCFR-STREP), contract number 012773 (FI6O), effective March 2005 - February 2009.

Improving Fuel Cycle Design and Safety Characteristics of a Gas Cooled Fast Reactor

Proefschrift

ter verkrijging van de graad van doctor
aan de Technische Universiteit Delft,
op gezag van de Rector Magnificus prof. dr. ir. J.T. Fokkema,
voorzitter van het College voor Promoties,
in het openbaar te verdedigen
op dinsdag 12 december 2006 om 12:30 uur

door:

Willem Frederik Geert VAN ROOIJEN
Natuurkundig ingenieur,
geboren te Haarlem

Dit proefschrift is goedgekeurd door de promotoren:
Prof. dr. ir. T.H.J.J. van der Hagen
Prof.[em] dr. ir. H. van Dam
Toegevoegd promotor:
Dr. ir. J.L. Kloosterman

Samenstelling promotiecommissie:

Rector Magnificus, voorzitter	
Prof. dr. ir. T.H.J.J. van der Hagen	Technische Universiteit Delft
Prof.[em] dr. ir. H. van Dam	Technische Universiteit Delft
Dr. ir. J.L. Kloosterman	Technische Universiteit Delft
Prof. W.M. Stacey, Ph.D.	Georgia Institute of Technology, USA
Prof. dr. M.J. van den Hoven	Technische Universiteit Delft
Prof. dr. ir. A.H.M. Verkooijen	Technische Universiteit Delft
Dr. G. Rimpault	CEA Cadarache, France

ISBN 1-58603-696-3

Keywords: Gas Cooled Fast Reactor, Generation IV Nuclear Reactor, Closed Fuel Cycle, Nuclide Perturbation Theory, Passive Safety, Lithium Injection Module

Published and distributed by IOS Press under the imprint Delft University Press

Publisher
IOS Press
Nieuwe Hemweg 6b
1013 BG Amsterdam
The Netherlands
tel: +31-20-688 3355
fax: +31-20-687 0019
email: info@iospress.nl
www.iospress.nl
www.dupress.nl

PRINTED IN THE NETHERLANDS

Contents

List of Figures

List of Tables

1 Introduction

This chapter gives a short overview of nuclear reactor types relevant to this thesis. After that, an overview is presented of the nuclear fuel cycle, the safety of fast reactors, and the Generation IV Initiative. The last section of the chapter is devoted to an explanation of the role of the Gas Cooled Fast Reactor within Generation IV, and an overview of the design choices and trade-offs that lead to the Generation IV GCFR design currently regarded as the reference design.

1.1 Nuclear reactor types relevant for this thesis

Liquid metal cooled fast breeder reactor

The only fissile material occurring in nature is uranium, which contains 0.7% of the fissile isotope ^{235}U and 99.3% of the non-fissile isotope ^{238}U. The non-fissile ^{238}U can be converted to the fissile isotope ^{239}Pu by neutron capture and subsequent decay ($^{238}\text{U} \longrightarrow [n, 2\beta^{-}] \longrightarrow {}^{239}\text{Pu}$), and ^{238}U is accordingly designated a *fertile* isotope. Conversion of fertile isotopes occurs in all nuclear reactors. If more than one fertile nucleus is converted for each fissioned nucleus, the net amount of fissile material in the reactor increases: the reactor is *breeding* fissile material. Given the fact that natural uranium contains 99.3% fertile material, breeding increases the effective use of the uranium resource. To sustain the nuclear chain reaction at least 1 neutron should be created per fission. To also convert one nucleus, at least one extra neutron is required. To maximize the possibility of conversion the number of new neutrons per fission should be maximized. The number of new neutrons generated by fission per neutron absorbed in a given nuclide is given by $\eta = \nu\sigma_f/\sigma_a$ [Duderstadt and Hamilton,

1976]. Here, ν is the number of neutrons released per fission, σ_f is the fission cross section of the nuclide and σ_a is the absorption cross section. All cross sections and ν are energy dependent, and thus η is also energy dependent. In figure 1.1 η is given as a function of the energy of the neutron causing fission. Fissions caused by high energy neutrons create a large number of new neutrons, allowing effective conversion of fertile material into fissile material. Reactors working on this principle are known as Fast Breeder Reactors (FBR, 'fast' because of the equivalence between energy and speed through $E = \frac{1}{2}mv^2$). As can be seen from the figure, Pu is a good FBR fuel, and breeding in a thermal spectrum is very difficult because of low neutron production.

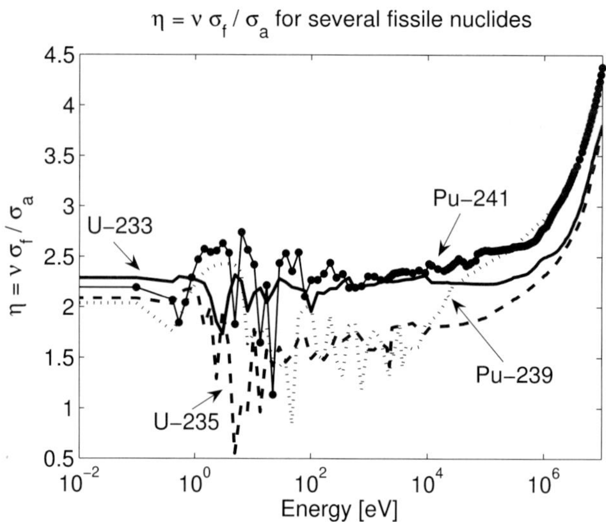

Figure 1.1. *Energy dependence of $\eta = \nu\sigma_f/\sigma_a$ for some fissile nuclides. In Fast Reactors the average energy of the neutrons causing fission is of the order of 100 keV, and plutonium would enable the highest neutron production. To enable breeding in a thermal spectrum reactor, ^{233}U is the best fuel.*

The production rate of new fuel is dictated by the capture rate in fertile material $N\sigma_c\phi = \Sigma_c\phi$, with N the atomic density of the capturing fertile nuclide, σ_c the capture cross section and ϕ the neutron flux impinging on the fertile material. To breed fuel quickly and efficiently, Σ_c and ϕ should be as high as possible. But to reach criticality Σ_c should not be too high, as $k_\infty \approx \frac{\nu\Sigma_f}{\Sigma_c+\Sigma_f}$, with Σ_f the macroscopic fission cross section . The contradictory requirement of large capture cross section and large fission cross section is usually solved by using a small driver core with a large fraction of fissile material, surrounded by blankets of fertile material. Neutrons leak from the driver core into the blankets where new fissile material is bred.

For economic operation, the specific power of the fuel, i.e. the power produced per unit mass of fuel, is the most important parameter. If the specific power is too low, there will be an excessive amount of nuclear fuel in the reactor (and the total fuel cycle) per produced unit of energy. Enrichment is also expensive, so a given amount of enriched fuel should produce as much energy as possible, preferably in a short time. This argument holds for any reactor, running in any possible fuel cycle scheme. To obtain good economic operation of the highly enriched fast reactor driving core, the power production should be maximized. The volumetric power production of nuclear fuel is given by $P = E_{\mathrm{rel}}\Sigma_f\phi$, where E_{rel} is the amount of energy released per fission, and the product $\Sigma_f\phi$ is the number of fissions per unit time per unit volume. A coolant is required to remove the power from the core, while maintaining the smallest possible volume fraction of coolant in the core and without introducing significant moderation of the neutrons. The classic solution is to use liquid metals to cool the core, like Na, NaK, PbBi etc. Hence the name traditionally given to this kind of reactor: Liquid Metal Fast Breeder Reactor (LMFBR).

Thermal spectrum Gas Cooled Reactors

In a Gas Cooled Reactor the heat generated from fissions in the fuel is transported out of the core by a gas. Moderation of the neutrons is provided by a moderator, which transfers a negligible amount of heat to the outside world. Graphite is commonly used as moderator. Gas cooled reactors usually use CO_2 as the coolant (e.g. MAGNOX and AGR reactors [Melese and Katz, 1984]), or helium (e.g. HTTR and HTR-10 [Lohnert, 2004, 2002]). A gas coolant has no phase change in the core, reducing the risk of boiling effects, and it can reach high outlet temperatures. Theoretically, the hot outlet gas could be run through a gas turbine to produce electricity in a Brayton cycle (direct cycle operation). Direct cycle operation requires a high outlet temperature (850 °C) to be efficient. At high temperatures, graphite is corroded by CO_2 ($C + CO_2 \longrightarrow 2\,CO$), which is the impetus for developing helium cooled reactors. Helium is chemically inert, allows high temperature operation, has a high specific heat and is neutronically transparent. Helium cooled reactors with high temperature operation are known as HTRs (High Temperature Reactor). HTR fuel is commonly in the form of small particles, the so-called TRISO fuel [Kugeler and Schulten, 1989, Massimo, 1976].
Because both graphite and helium have excellent nuclear properties, especially very low absorption, the volume fraction of fissile material in the core can be very low, and this means that even with a very low volumetric power density an adequate specific power can be obtained. Because of its low volumetric power density, the HTR is the reactor type with the highest level of inherent safety.

Gas Cooled Fast Reactors

Fast reactors can be cooled by gas. Even though most gaseous coolant candidates are light nuclei, the coolant does not introduce too much moderation because the number density of gas molecules is low. Coolants that have been considered include steam, CO_2 and helium [Melese and Katz, 1984]. The main advantages of gas coolant for a fast reactor are a reduction of the void effect (see section 1.3), reduction of parasitic absorption by the coolant, and a harder spectrum leading to better neutron economy and higher breeding gain. Lower enrichment is

possible while breeding more new fuel, increasing the economy of the reactor. Apart from that, most gas coolants are chemically less corrosive than Na or PbBi.
When GCFRs were first investigated in the late 1950s, the design choice was to use an adapted LMFBR core. The requirement of high power density in the core was maintained as rapid breeding of fissile material was the number one priority [Waltar and Reynolds, 1981]. To transfer the heat from the fuel pins into the coolant, fuel pins with roughened surface were required to enhance heat transfer to the coolant. The operating pressure of the reactors was high, between 7 and 15 MPa, to increase coolant density. The pressure drop of the coolant between core inlet and outlet was considerable, leading to the need of high power circulators to pump the gas through the system. GCFR safety poses special problems, and in the end, the economy of GCFR was never proven. A lack of both market demand and technology push lead to the extinction of most GCFR projects in the late 1970s.

1.2 The nuclear fuel cycle

Nearly all nuclear power reactors operational in the world are thermal spectrum reactors. These reactors use mainly UO_2 fuel, enriched in the fissile isotope ^{235}U (3% to 5%). In the enrichment process a large amount of depleted uranium is separated, for which there is currently not much use. During the irradiation of the fuel in a reactor, inevitably plutonium is produced, together with a small amount of heavier elements like Am, Cm etc, the so-called Minor Actinides (MA). After irradiation in the reactor and a subsequent cooling down period the spent nuclear fuel (SNF) contains some 5% (radioactive) fission products, 1% of fissile material, and 94% of ^{238}U. Currently, two options are available for the next step:

1. All material is conditioned for final storage. This is known as the Once Through Then Out (OTTO) fuel cycle.

2. The material is recycled to extract the re-usable materials, such as U and Pu. All other materials are conditioned for storage. This is known as the recycling or reprocessing fuel cycle.

Industrial scale reprocessing technology currently uses the PUREX process (e.g. [Long, 1978]), where uranium and plutonium are extracted from the SNF. All other materials are conditioned for final storage. The recycled Pu can be used to make new fuel, which is known as Mixed Oxide (MOx) fuel, chemical composition $UPuO_2$. Unfortunately, present-day reactors can only run safely if the MOx loading remains limited to some 30% of the total fuel inventory [Kloosterman and Bende, 1999]. State of the art PWR designs such as EPR and AP-1000 are able to use 100% MOx fuel, but there exists an increasing stockpile of plutonium from reprocessing. Also, the radiotoxicity of the ^{238}U left over from present-day enrichment plants will, in time (10^5 years), increase by 2 orders of magnitude [Hoffman and Stacey, 2003] due to radioactive daughter products. Although the same decay occurs in natural uranium ore, the depleted uranium is highly concentrated, and will form a large contribution to the future radiotoxicity due to present-day use of nuclear energy.

Nuclear proliferation issues

Nuclear material poses a proliferation risk: nuclear materials with a high fissile fraction can be used to make a nuclear weapon, and fertile materials can potentially be converted to fissile material. The 'quality' of a mixture of isotopes is usually judged by the fissile fraction. High quality materials are commonly referred to as 'Weapons Grade', as these materials can potentially be used without further steps to produce a weapon. Materials with a lower quality are designated 'Reactor Grade'. Plutonium has a high proliferation risk, as it can be produced with a high fissile fraction from ^{238}U in any nuclear reactor if the irradiation circumstances are chosen correctly. During the normal power operation of an LWR, Pu is produced with a lower quality isotopic composition (Reactor Grade), but a certain proliferation risk remains. In table 1.1 typical isotopic compositions are given for WG-plutonium and RG-plutonium. In the PUREX process plutonium is separated individually, and this is one of the reasons why some countries opt to not recycle SNF. In the blankets of a fast reactor, plutonium with a high fissile fraction is bred, which poses a proliferation risk. To reduce the risk of proliferation, reprocessing technologies without individual separation of actinides are an option. Reactors without blankets also increase proliferation resistance of the fuel cycle.

Table 1.1. *Isotopic composition (in weight %) of Weapons Grade plutonium, plutonium recycled one time, and recycled five times in an LWR. Data taken from Hesketh and Sartori [1999]. The quality of the Pu is usually measured by the fissile fraction (^{239}Pu + ^{241}Pu). Plutonium with a quality over 94% is designated Weapons Grade ([IAEA-TECDOC-1349, 2003]). Multiple recycling of plutonium in an LWR could increase proliferation resistance by degradation of the Pu-vector.*

	^{238}Pu	^{239}Pu	^{240}Pu	^{241}Pu	^{242}Pu
WG Pu	0.05	93.6	6.0	0.3	0.05
1st gen. MOx	1.8	59.0	23.0	12.2	5.0
5th gen. MOx	4.0	36.0	28.0	12.0	20.0

Partitioning and Transmutation

The radioactive fission products present in SNF generally have a short half-life compared to the MA, and as a result the long term (> 300 years) radioactive source term of the spent nuclear fuel is mainly due to the presence of plutonium and Minor Actinides. If the plutonium and MA can be removed from the SNF, the time during which the material remains hazardous can be greatly reduced [Stacey, 2001]. The separated MA could be stored in a geological repository. To reduce the need for such long-term storage, the MA could be destroyed by fission. Destruction of MA materials is most efficiently done using irradiation by fast (high energy) neutrons. For several MA isotopes the parameter $\alpha = \sigma_c/\sigma_f$ is given in figure 1.2. For high energies, α goes to zero: fission dominates over capture. In a high energy spectrum, the chance of fissioning a Minor Actinide nucleus is much larger than capturing a neutron, which would create an even heavier MA nucleus. The selective separation and subsequent destruction of Minor Actinides is commonly known as 'Partitioning and Transmutation' [Salvatores, 2005].

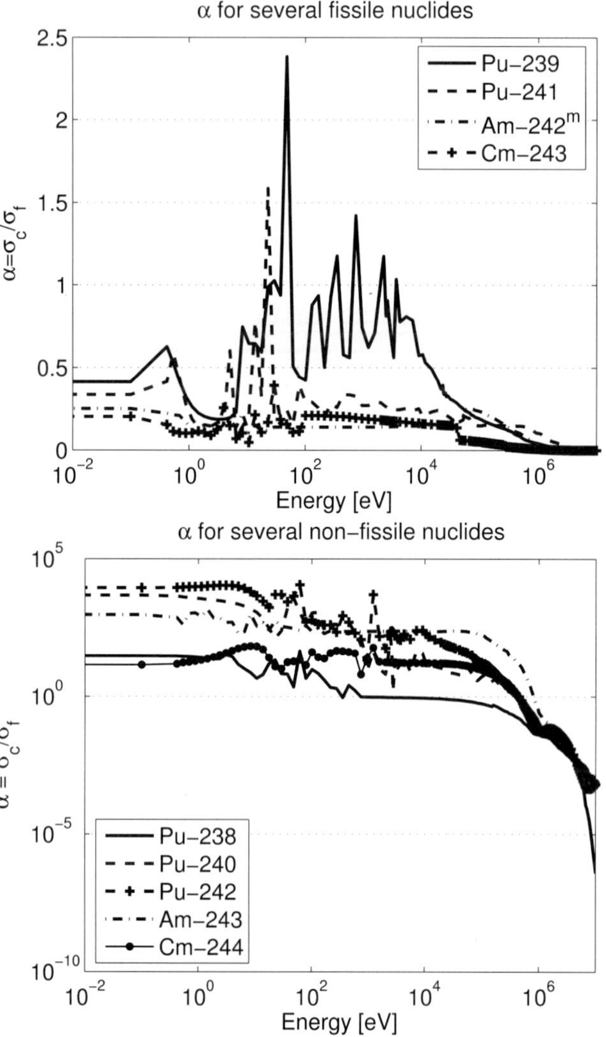

Figure 1.2. *The top graph shows the parameter $\alpha = \sigma_c/\sigma_f$ for several fissile isotopes, and in the bottom graph α is given for several non-fissile isotopes. Note the different scaling on the y-axis (top graph linear, bottom graph logarithmic). For all isotopes, α decreases with energy, with the most marked decrease for the non-fissile isotopes. In a fast neutron environment fission dominates over capture.*

The closed, sustainable nuclear fuel cycle

The increasing stockpile of depleted uranium from enrichment, the increasing amount of plutonium and its associated proliferation risk, together with the wish to partition and transmute MA to reduce the SNF lifetime, can all be addressed using fast reactors. A sustainable nuclear fuel cycle is characterized by the following aspects:

- In the reactor park enough new fissile material should be produced to allow refueling of the reactors, adding only fertile material to make the new fuel. In the current situation with an existing stockpile of plutonium, (rapid) breeding of fissile material is not necessary.
- Application of fast reactors to breed fissile material from fertile material, and to allow effective transmutation of currently existing stockpiles of plutonium and MA materials.
- An integral fuel cycle: the SNF is separated into one stream of fission products, and another stream of all Heavy Metals, without any separation of individual MA elements.

This kind of fuel cycle, where all Heavy Metals are recycled, and only relying only on natural fertile materials, is called the 'closed fuel cycle', as opposed to the OTTO and reprocessing fuel cycles which are both 'open ended'. Application of the closed fuel cycle has many advantages: potentially all uranium can be used to generate energy. It is shown in Cometto et al. [2004] that a closed fuel cycle with breeding in fast reactors can improve energy efficiency of the uranium resource 180 times over a simple LWR Once Through fuel cycle. Sustainability of nuclear power generation thus calls for the use of existing stockpiles of uranium as much as possible, while not producing new long-term hazardous material.

1.3 Safety of fast reactors

During steady state operation of a nuclear reactor there is a balance between the power produced in the core and the power transferred to the environment. If more power is produced than transferred, the temperature of the core increases with potentially severe effects. Two important situations are: (1) an increase of fission power caused by a reactivity insertion into the reactor, or (2) the situation where the fission chain reaction is stopped but there is no adequate coolant flow to transfer the heat of radioactive decay out of the core.
The reactivity ρ of a nuclear reactor is defined as $\rho = (k_{\text{eff}} - 1)/k_{\text{eff}}$. If the coolant density in a fast reactor is decreased, a couple of competing reactivity effects occur [Waltar and Reynolds, 1981]:

1. $+\rho$: spectral hardening due to a reduction of moderation by the coolant. In a harder spectrum, more neutrons are produced per fission and resonance absorption is decreased
2. $+\rho$: reduced absorption by the coolant

3. $-\rho$: reduction of scattering by the coolant leading to higher leakage of neutrons

4. $-\rho$: an absence of coolant leads to higher temperatures, which leads to larger absorption through the Doppler effect if a proper fuel composition is used (Doppler effect: increase of absorption caused by increased temperature of reactor materials)

To enhance the safety of fast reactors, the reactor geometry is chosen such that the third term (leakage) always dominates in case of coolant voiding. To enhance axial neutron leakage the core is made as flat as possible ('pancake core'). In GCFRs the coolant void effect is more benign than in liquid metal cooled reactors, because the coolant is transparent from a neutronic point of view [Melese and Katz, 1984].

In fast reactors special consideration should be given to reactivity introduction from geometrical changes. Core compaction, i.e. a rearrangement of the fuel into a more compact shape, may increase the reactivity of a fast reactor, unlike thermal spectrum reactors. Thermal reactors are designed to have an optimal fuel to moderator volume ratio, and a compaction of the core will lead to subcriticality due to a lack of moderation. In fast reactors on the other hand a more compact core with higher density is less 'transparent' from a neutronics point of view, and as a result the reactivity increases [Waltar and Reynolds, 1981].

In an operating nuclear reactor many radioactive fission products are formed. The radioactive decay of these fission products is the source of the so-called Decay Heat, which amounts to some 7% of the total power output of a nuclear reactor during steady state operation [Kloosterman, 2000]. If the fission chain reaction is stopped, the fission products will continue to decay, and the decay heat needs to be transported out of the reactor to avoid damage due to overheating. In a fast reactor core overheating may lead to core degradation, possibly followed by core compaction, with the associated risk of increased reactivity. This is known as the 're-criticality' problem.

Metal cooled fast reactors can take advantage of the high thermal conductivity of the coolant and natural convection to aid in Decay Heat Removal (DHR). This is especially true for pool-type reactors, which have a large pool of coolant to act as a primary heat sink. In the case of HTRs, their low core power density allows adequate DHR by conduction through the graphite to the environment. For Gas Cooled Fast Reactors, the safety case needs careful attention, especially a sudden depressurization of the primary system (Loss Of Coolant Accident, LOCA). As described above, the reactor operates at high pressure. A depressurization leads to a decrease of coolant density (coolant voiding), possibly introducing positive reactivity. Even if the fission chain reaction is adequately stopped, the high power density GCFR core cannot be cooled by conduction and natural convection alone. An adequate flow of coolant to extract the Decay Heat must be ensured at all times. In the past GCFR designs relied on an active Decay Heat Removal strategy, using (large) active backup circulators. The extra DHR precautions made GCFRs very intricate systems with a very doubtable safety characteristic [Chermanne and Burgsmüller, 1981, Torri and Buttemer, 1981].

1.4 The Generation IV initiative

The possible depletion of fossil fuel and the wish to limit CO_2 release into the atmosphere cause a new interest in nuclear energy as the only CO_2-free energy source with high capacity. There is a growing pressure from society to reduce the amount of long-lived nuclear rest material as far as possible, and to further increase the safety of nuclear power stations. These points are addressed in the Generation IV initiative. The Generation IV International Forum is an international research initiative for the fourth generation of nuclear power plants, envisaged to enter service halfway the 21st century [U.S. DOE Nuclear Energy Research Advisory Committee and the Generation IV International Forum, 2002]. Six reactor types have been selected for further research and evaluation in the Generation IV framework, the Gas Cooled Fast Reactor being one of the six concepts. Generation IV targets to improve all aspects of nuclear power generation: safety, economics, sustainability and availability. The six reactor types chosen all have their strong and weak points; used in a symbiotic system, the six reactor types should counterbalance their mutual weak points. For the reasons mentioned earlier, there is a renewed interest for fast reactors within Generation IV. Since (rapid) breeding is no longer the primary target for the fast reactors, a shift is made from high power density cores to fast reactors that will consume excess fertile material using reprocessed fuel from other reactors. The boundary conditions for a Gas Cooled Fast Reactor are thus improved. For completeness, the six Generation IV systems are listed here:

1. VHTR: Very High Temperature Reactor, with graphite moderator and helium coolant
2. SCWR: Supercritical Water Reactor
3. GCFR: Gas Cooled Fast Reactor, high temperature operation using helium coolant
4. SFR: Sodium cooled Fast Reactor
5. LFR: Lead cooled Fast Reactor
6. MSR: Molten Salt Reactor.

Within the Generation IV framework the GCFR concept targets sustainability, i.e. optimal use of resources while maintaining good safety and economical performance. The reference design of the Generation IV GCFR is:

- Fast reactor core, without fertile blankets, i.e. all new fissile fuel is bred in the core. In a fertile blanket, the fissile element is isotopically almost pure, which poses a proliferation concern.
- Breeding Gain equal to zero, breeding enough fissile material to refuel the same reactor adding only fertile material.
- The fuel specific power is low. By allowing a low specific power, the volumetric power density in the core remains limited. To make up for the economic penalty of low specific power, a highly efficient power conversion system with a direct coupled gas

turbine in a Brayton-cycle is the reference for electricity generation [Garnier et al., 2003].

- Helium is chosen as the reference coolant. To obtain a high efficiency in a Brayton-cycle using helium as the working fluid, a high reactor outlet temperature is necessary (850°C). To enable operation at such high temperatures, ceramics rather than steel are used as the structural material.
- The design of the core and fuel elements targets easy decay heat removal and exclusion of re-criticality, by using refractory (high melting point) materials, and by allowing a large coolant fraction in the core.

It should be noted that the application of ceramic materials in nuclear reactors has hitherto been limited to non-structural materials, like $UPuO_2$ fuel. In appendix B an overview is given of the mechanical and thermal properties of ceramics as a structural material.

The Generation IV GCFR

The choice of zero breeding gain and the absence of blankets determine the fuel composition: there is only a narrow band of possible isotopic compositions that will result in a zero breeding gain. The absence of blankets means that the reactor can be approximated by a simple homogeneous cylinder of a fuel/coolant mixture (neglecting the heterogeneity introduced by different fuel batches etc). For a homogeneous cylindrical reactor of a given volume, a *minimal* height to diameter (H/D) ratio exists, below which the reactor will never become critical due to excessive neutron leakage. On the other hand, an *optimal* H/D ratio also exists, where neutron losses by leakage are minimized. Now assume a homogeneous, cylindrical reactor of a given volume. If the coolant fraction is increased, the amount of fuel in the core is decreased, and thus the H/D-ratio must be chosen closer to the optimal value to obtain criticality. This is illustrated in figure 1.3, where the 'neutronic limit' is given as a function of the coolant fraction in the core. The neutronic limit is the *minimum* value of H/D to obtain criticality with a given fraction of coolant in the core.
There is also a thermohydraulic limit on the H/D ratio of the reactor. If, for the same power density and reactor volume, the H/D ratio becomes larger (i.e. increase core height H, reduce diameter D), more power has to be transferred per coolant channel. This requires a larger coolant mass flow per coolant channel, resulting in a larger pressure drop Δp_{core} over the core. This means that for a given fraction of coolant in the core, there is a *maximum* value of H/D to stay below a given core pressure drop. This is illustrated in figure 1.3 by the 'TH limit'.
A reactor can only be designed if the neutronic limit of H/D is smaller than the TH limit, like in figure 1.3. Note that the neutronic limit is not necessarily smaller than the TH limit, e.g. the TH limit may result in a H/D value too low to obtain criticality. In practice, choosing a low power density will enlarge the region of possibilities: lower power density means that it is easier to stay within TH-limits, and at the same time the reactor volume will be larger for the same power output, giving larger neutronic margins by reduction of leakage. Ultimately

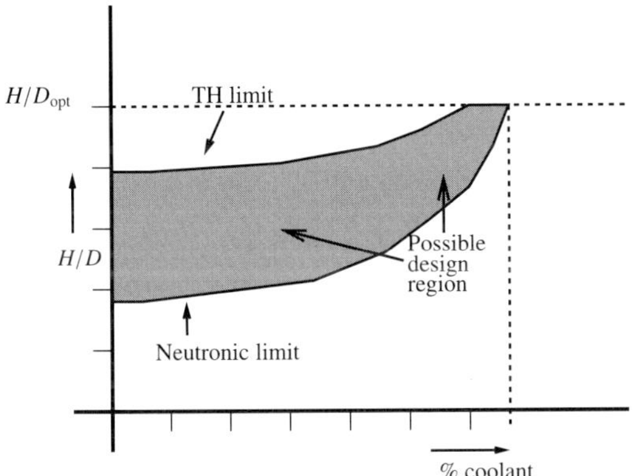

Figure 1.3. *Schematic illustration showing a region of feasible GCFR designs bounded by neutronic and thermohydraulic (TH) limits. The x-axis gives the percentage coolant in the core. The neutronic limit is the lowest H/D ratio which will allow criticality for a chosen fuel composition and coolant fraction. The TH limit gives the largest H/D ratio allowing operation below a given Δp_{core}. A design is only possible if the neutronic limit is below the TH limit, like in this figure.*

the power density has to be chosen to yield a reasonable specific power, i.e. power per unit mass of fuel in the core. The limiting factors are the amount of fuel material allowed in the fuel cycle per unit energy produced, and the availability of fissile materials versus the required energy output. Having a large inventory of nuclear materials for a low amount of energy produced is bad for economy and proliferation resistance.

1.5 Contents of this thesis

In chapter 2 of this thesis an overview is given of GCFR developments from the 1950s until the early 1980s. In the same chapter, the possible future of GCFR development is sketched, and an overview is given of the GCFR concepts studied within the Generation IV framework. In the other chapters, the following subjects are investigated:

- The possibility of obtaining the closed fuel cycle (see section 1.2) is addressed in chapters 3 to 5. The possibility of obtaining a closed fuel cycle is not apparent a priori (the fuel composition necessary for self-breeding may preclude a critical system, for instance). It is shown that the closed fuel cycle can be obtained if the reprocessing of the spent fuel is efficient enough. It will be shown that homogeneous reactors (i.e. without blankets) intended to have self-breeding have the potential for very low burnup reactivity swing, and that in fact a positive reactivity swing may occur for certain fuel compositions.
- Chapter 3 treats the design of a 2400 MWth GCFR using coated particle fuel. It will be

shown that a pebble bed type of core is not possible for GCFR. On the other hand, as will be shown, HTR TRISO coated particle fuel can be redesigned for application in a GCFR. The reactor is capable of obtaining a closed fuel cycle. In the chapter, two designs for the coated particle fuel are proposed. This chapter also treats the performance of a plate-type reactor with a fuel containing additional Minor Actinides.

- In chapter 4 the subject of Breeding Gain is treated theoretically. A new definition of reactivity weights is proposed, alongside a new definition of Breeding Gain. First order perturbation theory is then developed to estimate the effect on the performance of the closed fuel cycle due to small variations of the initial fuel composition.
- In chapter 5 the theoretical framework is applied to a 600 MWth GCFR in a closed fuel cycle. It is shown that the new definition of reactivity weights is consistent with existing definitions, and that the new definition of BG does give $BG = 0$ if the multiplication factor k_{eff} is the same from cycle to cycle. It is shown that adding a small amount of MA to the fuel is beneficial for BG, and that first order perturbation theory provides a powerful tool to estimate the effect of changes to the fuel and fuel cycle.
- Chapter 6 treats the design of a passive reactivity control device intended to control the reactor power and temperatures under accidental conditions when other control elements have failed. The proposed passive device is capable of protecting the fuel against excessive temperatures.

This thesis closes with a chapter summarizing the most important conclusions, and with two appendices: appendix A describes a transmutation code especially written in the scope of this thesis research. In appendix B the (thermo)mechanical properties of ceramics as a structural material are discussed.

2

History of GCFR development

The possibility to convert non-fissile nuclides into fissile ones by neutron capture was recognized from the earliest stages of nuclear reactor development. In the 1940s and 1950s, the available amount of natural uranium was expected to be insufficient to fuel a large number of power reactors. This led to a feverish development of 'high energy neutron nuclear reactors', also known as 'fast reactors', intended to breed fissile isotopes from fertile material: ^{239}Pu from ^{238}U, ^{233}U from ^{232}Th.

The first fast reactor, called 'Clementine', was built at Los Alamos (US) and reached criticality in late 1946. In parallel, the Experimental Breeder Reactor (EBR) was built by Argonne National Laboratory, from the start conceived as a prototype of a power reactor. EBR went critical in mid 1951. From the beginning of the fast reactor programme it was recognized that fast breeder reactors, in order to meet the breeding targets and obtain a reasonable specific power, needed to be systems with a very high power density. To accommodate the high power density, liquid metals were selected as coolant. The first reactors used liquid metals like Hg and NaK, and in due course sodium became the reference coolant for what was dubbed the 'Liquid Metal Fast Breeder Reactor (LMFBR)'.

Gas cooling is interesting for fast reactors because it reduces the risk of reactivity induced transients due to coolant voiding. A gaseous coolant also reduces parasitic absorption by the coolant, resulting in better neutron economy and improved breeding gain. On the other hand the high power density required for economical operation of a breeder reactor makes gas cooling a challenging problem. In the 1950s much progress was made in thermal gas cooled reactor programs, like the development of helium as a coolant, and elevated temperature operation to improve the thermodynamic efficiency of the power conversion system. From 1962 onwards several Gas Cooled Fast Reactors were proposed. All of these reactors were based on LMFBR designs with slight changes for the specific demands of gas cooling. The main difference with LMFBR fuel is that GCFR fuel pins were roughened with small ridges perpendicular to the flow direction to induce turbulence and improve heat exchange. High pressure operation (7 - 15 MPa) is required to achieve an adequate coolant density. The high pressure requires a special reactor housing, with almost all concepts relying on a primary system in concrete, the so-called Prestressed Concrete Reactor Vessel (PCRV), which was under development for thermal HTR applications.

In this chapter, an overview is given of the various GCFR programs initiated in the 1960s and 1970s by all major nuclear industries and research organizations. This overview is by no means exhaustive, but it serves to give an illustration of what had already been achieved long before the Generation IV program was initiated. The last section of the chapter is devoted to the GCFR design proposed for Generation IV.

2.1 Germany: the Gas Breeder Memorandum

In Germany the nuclear research establishments at Karlsruhe and Jülich, together with partners from industry, prepared a document on Gas Breeder Reactors known as the Gas Breeder Memorandum ('Gasbrüter-Memorandum', 1969) [Dalle Donne and Goetzmann, 1974]. This memorandum defined 3 reactor concepts, all featuring helium cooling. Steam and CO_2 were reviewed as coolant candidates but deemed inadequate. In the Gas Breeder Memorandum the main focus was on a conventional core (fuel assemblies extrapolated from an LMFBR design, PCRV extrapolated from thermal HTR), with pin-type fuel, stainless steel cladding and a secondary steam cycle, with all blowers and steam generators integrated into the PCRV. Limited research was done into a direct-cycle reactor, and into coated particle fuel, although the two were not necessarily combined into one concept. Some design data can be found in Dalle Donne and Goetzmann [1974]. The plans for this reactor reached a level of detail, with material irradiations planned in the BR-2 reactor in Mol (Belgium) in the mid-seventies. The design is interesting because it already emphasized the need of keeping an elevated backup pressure (2 to 3 bar overpressure) around the primary system after a LOCA in order to cool the core.

2.2 US: General Atomics

In the U.S. General Atomics announced plans for a Gas Cooled Fast Reactor in 1962. GA prepared designs for a 300 MWe demonstration plant and a 1000 MWe commercial plant [Waltar and Reynolds, 1981]. In 1968 the GCFR Utility Program was started to design, license and build a 300 MWe demonstration plant [Simon et al., 1974]. In 1973 the target was set for the GCFR to start operation in 1983. The main parameters of the 300 MWe GCFR are given in table 2.1, and an illustration of the plant layout is given in figure 2.1 (taken from Simon et al. [1974]). The GA design has helium coolant and UO_2 fuel in stainless steel cladding. The entire core is based on LMFBR technology with slight adjustments for the gaseous coolant. The fuel pins are roughened to enhance heat exchange. The primary system is housed in a PCRV, into which all blowers and steam generators are integrated. A last reference to the GA GCFR demonstration design was found in 1981, when the power output was increased to 350 MWe, but the safety case remained problematic [Torri and Buttemer, 1981].

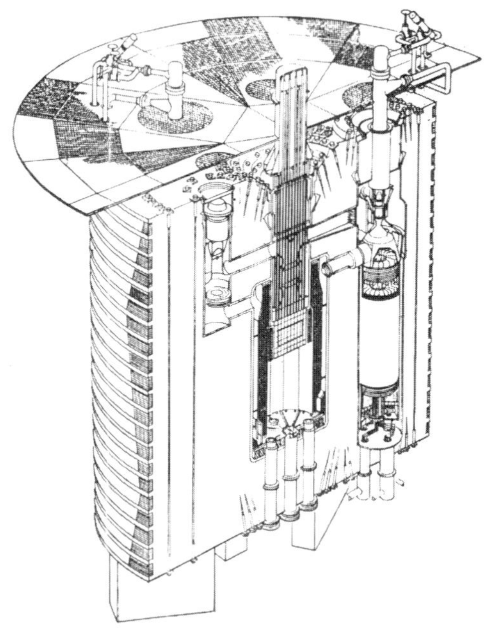

Figure 2.1. *Primary system layout for the General Atomics GCFR Demonstration plant. The core cavity is shown empty. The overall system layout is typical for all gas cooled reactor designs of the era. The large cavity on the right side of the core contains a boiler and a blower. The smaller cavity on the left side houses one of the active DHR systems. Figure reproduced from Simon et al. [1974].*

In the US a renewed interest exists for GCFR within the Generation IV framework. Current US designs include a Pebble Bed GCFR (PB-GCFR, [Taiwo et al., 2003]), as well as designs based on prismatic blocks ([Handwerk et al., 2006], also see section 2.6).

2.3 Europe: the Gas Breeder Reactor Association

In Europe a number of players in the nuclear field joined forces to develop a gas cooled fast reactor: the Gas Breeder Reactor Association. This group proposed a first design (GBR-1) in 1970, a 1000 MWe reactor featuring helium coolant, pin-type fuel, conventional outlet temperature and a secondary steam cycle. This design was followed by GBR-2 and -3 (1971), also 1000 MWe reactors but using coated particle fuel, slightly elevated outlet temperature,

Table 2.1. *Design data for four early GCFR proposals. The small General Atomics GCFR was envisaged as a prototype, the GBRs were envisaged as large scale commercial systems. From the three GBR concepts, the GBR-4 was adopted as the reference design.*

Reactor	GA GCFR	GBR-2	GBR-3	GBR-4
Coolant	He	He	CO_2	He
Thermal power [MW]	835	3000	3000	3450
Fuel type	pins	particle	particle	pins
Fuel material	$UPuO_2$	$UPuO_2$	$UPuO_2$	$UPuO_2$
$T_{core,in}$[°C]	323	260	260	260
$T_{core,out}$[°C]	550	700	650	560
Pressure [MPa]	8.5	12.0	6.0	12.0
Pressure drop [MPa][a]	0.37	0.34	-	0.24
Core height [m]	1.0	1.0	1.0	1.4
Core diameter [m]	2.0	-	-	-
Power density [MW/m^3]	-	-	-	-
Breeding Gain	0.4	0.43	0.36	0.42
Year of design data	1974	1972	1972	1974

[a]Pressure drop over primary circuit for GA design, core only for GBR designs

and helium coolant (GBR-2) resp. CO_2 coolant (GBR-3) [Chermanne, 1972b]. The 3 designs finally culminated in GBR-4, a 1200 MWe reactor with helium cooling and pin-type fuel. Table 2.1 lists the main design data for GBR-2, -3 and -4. Like other designs of the era, the core, blowers and steam generators were integrated into a PCRV. A cross-section of the GBR-4 PCRV is shown in figure 2.2.

For GBR-2 and GBR-3 detailed designs were prepared of the coated particles, and two designs were proposed for the fuel assemblies to hold the coated particles. The fuel assemblies for GBR-2 (helium coolant) and GBR-3 (CO_2 coolant) are illustrated in figure 2.3. For GBR-2, each fuel assembly consists of 7 fuel cylinders. Each fuel cylinder consists of 2 perforated concentric annuli with coated particles packed between them. Helium flows inward to keep a compressive stress on the inner tube. The inner tube would have been made in SiC, while other parts would be stainless steel.

The GBR-3 assembly consists of a 'stack of saucers'. The coolant flows up through the central cylinder, then flows radially through the bed of coated particles, then flows up and out of the core. The cold parts are made of steel, hot parts of SiC. The fuel assembly for the GBR-4 design is less ambitious, and is based on an LMFBR fuel assembly with pin fuel. An overview of the GBR-4 fuel assembly features is given in figure 2.4, because it is very typical of all GCFR fuel pins designs of that time. Each fuel pin holds a number of traditional MOx pellets. Surface roughening enhances turbulence and heat transfer. The high helium velocity requires many restraining devices to prevent the fuel pins from vibrating too violently. Spacer wires, traditionally used in fast reactors, are not strong enough. Thus grid spacers are used, which have a very intricate design to be strong enough and not introduce too much drag.

The GBR-2 design is interesting because it resurfaces in modern design proposals for GCFRs,

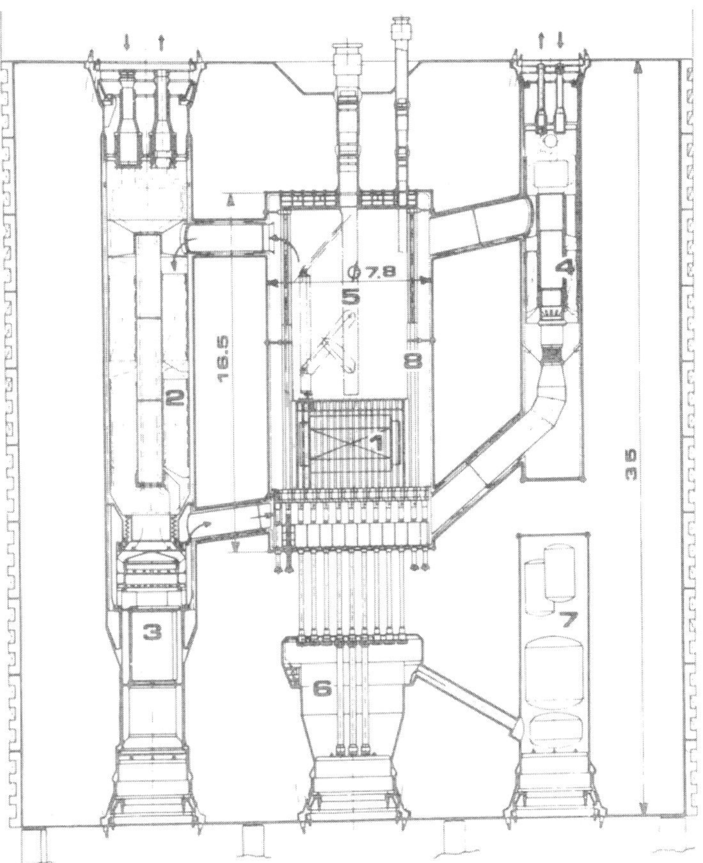

Figure 2.2. Vessel cross section of the PCRV for GBR-4 (1200 MWe). All dimensions in m. 1: core, 2: main steam generator, 3: main gas circulator, 4: emergency cooling loop, 5: fuel manipulator, 6: control cavity, 7: helium purification plant, 8: neutron shield. Notice the design of the fuel manipulator. The holes in the PCRV are kept as small as possible, so to change the fuel assemblies, this kind of fuel handling machine is used in virtually all GCFR designs (cf. figure 6.1). To be able to move the fuel assemblies around, the height above the core is at least as high as the fuel assemblies.

for instance in Japan [Konomura et al., 2003]. The objective of the coated particle fuels was to increase the coolant exhaust temperature to improve the thermodynamic efficiency of the secondary steam cycle. For both GBR-2 and -3 coated particles were only used for the driver fuel, the blankets employed traditional pin-type fuel. This solution was chosen because at the time of design, reprocessing of coated particle fuel was not proven. GBR-2 and -3 required several ceramic parts, most notably the structures at the outlet side. The fabrication

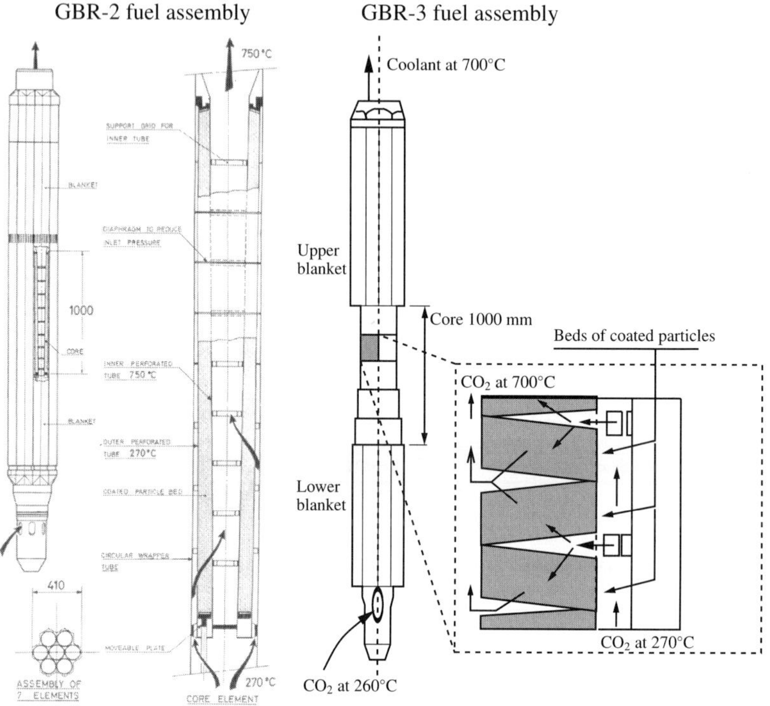

Figure 2.3. *Fuel assemblies for GBR-2 (left) and GBR-3 (right). See the text for the explanation.*

difficulties related to large ceramic parts led to the development of GBR-4, which is a much more conventional design. In GBR-4, the outlet temperatures are decreased, enabling the use of stainless steel components throughout the core. The plant efficiency is lower, which is offset by a larger total output of the reactor: from 1000 MWe to 1200 MWe. A last reference to the GBR-4 design was found in Chermanne and Burgsmüller [1981], where the safety case for large GCFR cores is discussed.

2.4 The Soviet Union: dissociating coolant

In the Soviet Union a GCFR programme was initiated focusing on an organic coolant: N_2O_4. In the core the N_2O_4 would dissociate through two endothermic chemical reactions [Melese and Katz, 1984]:

$$N_2O_4 \rightleftharpoons 2\,NO_2 \rightleftharpoons N_2 + 2\,O_2$$

Operating temperature was comparable to those of other contemporary GCFR designs, with

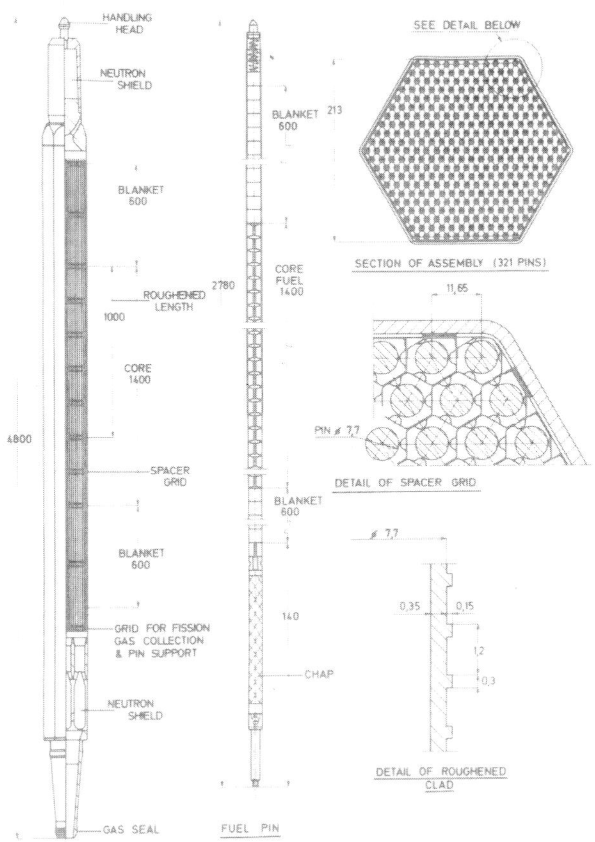

Figure 2.4. *Overview of the GBR-4 fuel assembly. An example of roughened cladding is in the lower right figure. Note the grid spacers: sturdy wires are woven around the pins, and then tightened.*

a somewhat higher pressure (values between 16 MPa and 25 MPa were found). The major advantage of the organic coolant lies in the possibility of condensing the working fluid in the heat exchanger, thereby greatly reducing the pumping power. The system operates much like a refrigerator. Also the combined effects of evaporation and a chemical reaction absorb a large amount of heat from the core, so the mass flow of coolant can be relatively small. The organic coolant is very corrosive. This problem was solved by the development of chromium dispersion fuel pins in the late 1970s (small inclusions of U metal or UO_2 in a matrix of chromium) and extensive research into the corrosion behavior of various steel types [Nesterenko et al., 1983]. This last paper also mentions irradiation experiments on the chromium dispersion fuel pins in a test rig using N_2O_4. As with other GCFR programs, no references are found later than the early 80s.

2.5 UK: ETGBR/EGCR

In the late 1970s a U.K. program was initiated into an 'Existing Technology Gas Breeder Reactor' (ETGBR). This design focused on joining the experience gained in the U.K. on sodium systems (PFR, Dounreay) and the thermal, CO_2-cooled AGR reactors. The fuel assemblies used stainless steel cladding with surface roughening, while the entire system was to be housed in a concrete vessel as used for the AGRs. ETGBR used CO_2 coolant, and had a lower power density than LMFBRs, with the expected higher breeding gain to make up for the difference [Kemmish et al., 1981]. The ETGBR is not very different from other designs of the same era for GCFRs. However, the ETGBR idea lingered on for a long time, well into the late 1990s. At that late stage, the ETGBR was rebranded EGCR: Enhanced Gas Cooled Reactor. EGCR was proposed as an actinide burner, first within the EFR (European Fast Reactor) program, and later in the CAPRA/CADRA study [Sunderland et al., 1999]. By then the reactor featured 3600 MWth, CO_2 cooling, and nitride fuel in fuel pins.

2.6 Japan: prismatic fuel

In Japan a fast reactor programme was initiated in the 1960s, including sodium and gas cooled reactor concepts. Kawasaki Heavy Industries (KHI) investigated GCFR concepts cooled with steam, CO_2 and helium [Mochizuki et al., 1972]. The helium concept was based on LMFBR technology, but KHI opted for a very low core, to reduce the pumping power requirements. The flat core also increases breeding gain but requires a larger fissile fraction. Investigations into the GCFR concept seem to have continued uninterruptedly in Japan, culminating in the late 1990s in a GCFR design proposal by JNC. This reactor also features a core with a low height/diameter ratio ('pancake core'), and uses coated particle fuel. A nitride fuel compound is chosen for the kernels. Buffer layer and sealing layers are made of TiN instead of graphite and SiC. Two types of fuel assemblies are proposed. One fuel assembly resembles that of GBR-2: coated particles are arranged in an annular bed, with the helium flowing radially through the bed [Konomura et al., 2003]. The other proposal has large prismatic blocks, filled with a mixture of coated particles and a matrix material (TiN, SiC or ZrC). Coolant channels run axially through the blocks. All structural parts are made in SiC. Thermal output is 2400 MWth, with a high power density of 100 MW/m^3. The coolant is helium and a direct cycle energy conversion system is envisaged. In the scope of the Generation IV initiative, JAEA is researching a GCFR based on prismatic fuel blocks containing a mix of coated fuel particles and a SiC matrix. An illustration of the prismatic fuel block is given in figure 2.5.

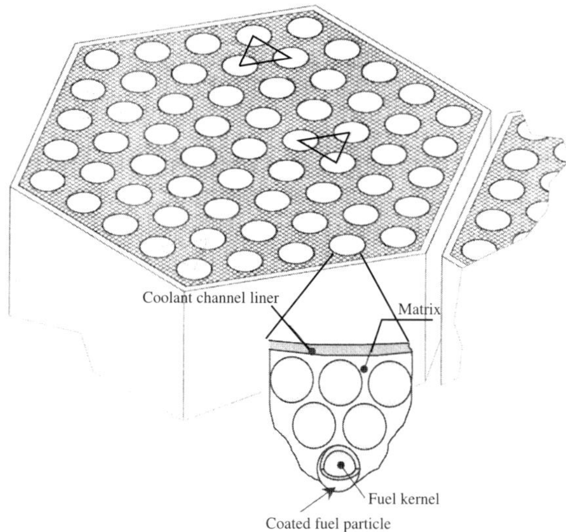

Figure 2.5. *Illustration of the prismatic fuel block proposed by JAEA for a GCFR. The coated fuel particles are embedded in a matrix material, which has two main tasks: transport heat from the fuel particles to the coolant channels, and structural rigidity. To increase the volume fraction of coated fuel particles, particles with different diameters may be used.*

2.7 The future of GCFR

GCFR designs of the past were all derived from sodium cooled reactors, with the main focus on improving the neutron economy. Neutron economy was an important issue in those days because of the expected lack of nuclear fuel in the (then) foreseeable future [Waltar and Reynolds, 1981]. However, the nuclear industry did not develop according to the expectations, and the expected shortage of uranium never occurred. Many fast reactor programs were canceled, due to a lack of demand from the market and a lack of fundamental improvements in sodium reactor technology: sodium reactors remained expensive and complicated installations, while LWRs could produce electricity at low cost.

Within the Generation IV initiative there is a role for GCFR as a sustainable reactor, consuming excess fertile material, while not producing extra MA waste. To limit the risk of releases of radioactive materials into the environment, matrix or coated particle fuel is preferred over pin fuel. The Decay Heat Removal strategy relies on natural circulation where possible. To limit the risk of depressurization, the primary circuit is enclosed in a second containment ('close containment' concept).

A large coolant faction is chosen in the core. This not only reduces core pressure drop during normal operation, it also increases the potential of decay heat removal using natural circulation. To obtain criticality with a large fraction of coolant in the core, carbide fuels are selected because of their high density of heavy metal. Plate-type fuel is preferred over pin-type fuel, to reduce coolant friction. As an illustration, figure 2.6 gives an overview of the reactor core of a Generation IV 2400 MWth GCFR with direct-cycle operation.

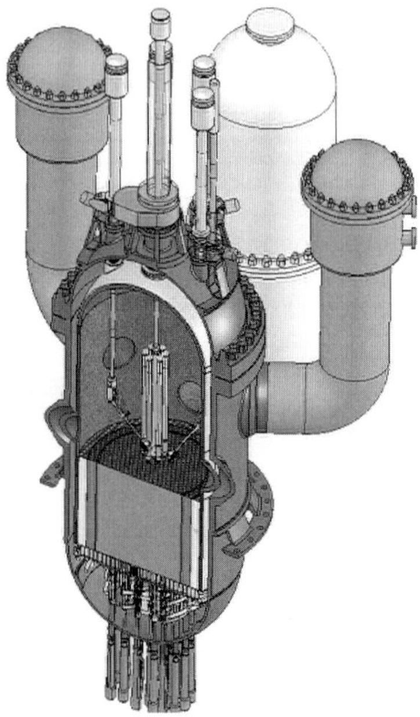

Figure 2.6. *Generation IV 2400 MWth GCFR (CEA design). The large light-colored structure houses one of the three turbomachines (3 x 800 MWth). The other two vessels house two of the three DHR circuits (heat exchangers and blowers). Above the core the fuel handling machine is visible. The entire primary system, including turbomachines and DHR-loops is to be housed inside a second, close containment, inside which an elevated backup pressure is to be maintained under accidental situations.*

For completeness, a small overview is presented in table 2.2 of the GCFR types proposed at the time of writing of this thesis (mid-2006). In Europe a small scale prototype reactor (ETDR, Experimental Technology Demonstration Reactor) is under investigation (CEA / Euratom). It is a prototype GCFR, intended to test and qualify materials and codes for Generation IV GCFR designs. It will be started with a conventional core, using oxide fuel in stainless steel cladding. The core will be gradually converted to use the ceramic fuel elements intended for Gen IV GCFR. Seven proposals for large GCFRs are under investigation by several research institutions within the Generation IV framework. Two main variants are adopted: a 600 MWth variant and a 2400 MWth variant. Direct and indirect cycle Power Conversion Systems (PCS) are considered. As far as fuel options are concerned, the priority is given to ceramic plate fuel containing a ceramic matrix fuel. The plate type fuel is a challenging design, and therefore pin-type fuel is maintained as a backup.

Table 2.2. *Design data for several Generation IV GCFRs, per July 2006. For all Direct Cycle reactors, helium is the preferred coolant. For the GFR600 Indirect Cycle variant, supercritical CO_2 (S-CO_2) is the secondary coolant. For the 2400 MWth Indirect Cycle concept, various secondary cycles and working fluids are under consideration. The JAEA GCFR has blankets to breed just enough new fissile material to allow operation in a closed fuel cycle.*

Concept	ETDR	GFR600	GFR600	GFR2400	GFR2400	JAEA GFR
Power [MWth]	50	600	600	2400	2400	2400
PCS	-	direct	indirect	(in)direct	direct	direct
Coolant	He	He	He/S-CO_2	He	He	He
Power density [MW/m^3]	100	103	103	100	100	90
Specific power [W/gHM]	-	45	45	-	42	36
$T_{core,in}$[°C]	250	480	≈ 400	480	480	460
$T_{core,out}$[°C]	525	850	≈ 625	850	850	850
Core H/D	0.86/0.86	1.95/1.95	1.95/1.95	1.55/4.44	1.34/4.77	0.9/5.9
Pressure [MPa]	7.0	7.0	7.0	7.0	7.0	7.0
Fuel type	pins[a]	plates	plates	plates	pins	blocks
Fuel mat.	$UPuO_2$	UPuC	UPuC	UPuC	UPuC	UPuN
Struct. mat.	AIM1[b]	SiC	SiC	SiC	SiC	SiC
Refl. mat.	AIM1	Zr_3Si_2	Zr_3Si_2	Zr_3Si_2	Zr_3Si_2	SiC
Vol.% c/s/f[c]	-	55/20/25	55/20/25	40/37.6/22.4	55/23/22	25/55/20
BG	-	-5%	-5%	-5%	0	0.03/0.11[d]
Design year	2004	2006	2006	2004	2006	2006

[a] Plate fuel in later stage
[b] Austenitic Improved Material 1, a variation of 15/15Ti stainless steel (SS-316), developed especially for fast reactors (Phénix and SuperPhénix)
[c] Coolant, structural materials, fuel
[d] Without / with blanket

3

Burnup and fuel cycle study for a GCFR using Coated Particle fuel

This chapter gives the results of the first investigations into the Generation IV Gas Cooled Fast Reactor in the scope of this thesis work. Investigations are carried out identifying the possible chemical fuel compounds, fuel form, core layout and reprocessing options compatible with the Generation IV objectives. The work leads to a design proposal for a 2400 MWth GCFR with a power density of 50 MW/m^3, using coated particle fuel. A burnup study is performed to estimate the performance of the proposed core concept in the closed fuel cycle. The results were reported in an article by Van Rooijen et al. [2005], upon which the first part of the chapter is based. The second part of this chapter concerns a burnup study performed in the scope of the European 6th Framework Program (FP6) GCFR-STREP (*S*pecific *T*argeted *RE*search *P*roject). The STREP investigates GFR600, a 600 MWth GCFR with plate-type fuel. An overview of some relevant results is presented.

3.1 Fuel for a Gas Cooled Fast Reactor

Depending on the type and operating conditions of a nuclear reactor, the nuclear fuel can occur in various geometrical shapes and various chemical compounds. In table 3.1 an overview is given of some basic properties of candidate fuel compounds for the Generation IV Gas Cooled Fast Reactor.

Oxide fuel ($UPuO_2$) is currently the standard fuel compound for all nuclear power reactors, both thermal and fast systems. Carbide and nitride fuel have a much higher density of Heavy

Table 3.1. *Property data for the candidate fuel compounds for Generation IV GCFR. Oxide is the current reference fuel compound for thermal and fast reactors. For Generation IV reactors, carbide or nitride fuel is preferred because of the high density of Heavy Metal, and high thermal conductivity.*

	$UPuO_2$	UPuC	UPuN
Density [g/cm^3]	11.0	13.6	14.3
HM Density [g/cm^3]	9.7	12.9	13.5
T_{melt} [°C]	2775	2480	2780
Thermal conductivity [W/mK]	2.9	19.6	19.8
PUREX compatible	yes	no	yes

Metal (HM, nitride almost 40% higher than oxide), and much higher thermal conductivity. However, carbide fuel is not compatible with PUREX reprocessing, not compatible with water and pyrophorus [Bailly et al., 1999]. Nitride fuel is not used because of the ^{14}C production through the $^{14}\text{N} \longrightarrow [(n,p)] \longrightarrow {}^{14}\text{C}$ reaction. This problem is solved if the nitrogen is enriched in ^{15}N, which has a natural abundance of 0.37%. For these reasons, carbide and nitride fuel are not used commonly.
For the Gas Cooled Fast Reactor, there are 3 basic fuel forms to choose from:

1. Pins
2. Plates (matrix fuel)
3. Fuel particles (particles may or may not be embedded in a matrix)

Pin-type fuel with metallic cladding is the standard fuel form for power reactors. The fuel is in the form of pellets. Gaseous and volatile fission products may escape from the fuel pellets, and are trapped in a fission gas plenum in the pin. Because fission products can escape easily from the pellets, pin-clad pellet fuel is only considered as a backup solution for Generation IV reactors.
Plate fuel is commonly used in research reactors. The fuel is usually a matrix type fuel. The fuel matrix can be an alloy of U and/or Pu metal, embedded in a metallic matrix (METMET). Other possibilities are a ceramic fuel compound embedded in a metallic matrix (CERMET), or a ceramic fuel compound embedded in a ceramic matrix (CERCER). For example, the HOR at Delft University had Highly Enriched Uranium fuel elements with an Al-U alloy surrounded with Al (METMET), and nowadays uses U_3Si_2 in an Al matrix (CERMET) [Verkooijen and de Vries, 2004]. Matrix fuels can reach high burnup, and fission products remain trapped in the matrix.
Particle fuel is commonly used in thermal, high temperature gas cooled reactors. Fission products can be released from the fuel, but remain trapped within the particle. The particles themselves can be surrounded by a matrix to make larger fuel elements (pebbles or compacts). For Generation IV reactors matrix or particle fuel are the preferred fuel form.

3.2 Reprocessing strategies

For Generation IV a closed nuclear fuel cycle is envisaged, which requires reprocessing of the spent nuclear fuel. Reprocessing is necessary for several reasons:

- Fission products are parasitic absorbers, deteriorating for instance safety performance of the reactor
- In a nuclear reactor, all atoms of the reactor materials (cladding, fuel, moderator, coolant) are interacting with neutrons. This interaction causes damage in the materials, for instance due to inelastic scattering, where energy is transferred from the neutron to an atom in a structural material. The atom can be removed from the lattice site it occupies in the crystal, damaging the crystal structure. As a result degradation of materials occurs during irradiation. If for instance the properties of the fuel become too degraded, safe operation of the reactor may not be possible. Thus at some point the fuel (and other materials in the reactor) have to be replaced.
- Gaseous and volatile fission products need to be removed from the fuel.

Several chemical processes exist to perform reprocessing. Only one of these processes, the PUREX (*P*lutonium *U*ranium *R*eduction and *EX*traction) process, is available on an industrial scale. In the PUREX process [Bailly et al., 1999, Long, 1978], the spent nuclear fuel (SNF) is dissolved in HNO_3, after which first the uranium is extracted, and subsequently the plutonium. All other chemical elements remain in the solution, and are prepared for final storage. PUREX does not allow a closed nuclear fuel cycle, because neptunium, americium and curium are not recycled, but instead remain in the solution with all fission products. PUREX is compatible with oxide and nitride fuel.
In the scope of Generation IV and the closed fuel cycle the most interesting technology is pyrochemical processing with electrorefining. In this process, the SNF is dissolved in a chloride salt at high temperature. Electrolysis provides a separation between actinides and fission products. Actinides are not separated individually in this process [Long, 1978, Ackermann et al., 1997, Walters, 1999]. Another pyrochemical process is AIROX (*A*tomics *I*nternational *R*eduction *OX*idation), where the SNF is reacted with oxygen, and due to the volume increase from UO_2 to U_3O_8 the SNF is pulverized. By heating, gaseous (Kr, Xe) and volatile fission products (Cs, Rb, Te, I) are removed, all other constituents remain in the fuel. The U_3O_8 powder is reduced to UO_2 to make new pellets [Long, 1978, Choi and Park, 2006, Ikegami, 2005]. The resulting fuel contains both actinides and fission products. Reprocessing where all actinides are recycled indiscriminately is commonly referred to *integral recycling*.

3.3 Coated particle fuel for a Gas Cooled Fast Reactor

The Generation IV GCFR should not have blankets to breed fissile fuel (i.e. all breeding takes place 'in-core'). This implies a rather low fissile fraction in the fuel. The requirements on steady state pressure drop and natural circulation under accidental situation dictate a rather large fraction of coolant in the core. The tentative volume fraction of fuel, structural materials and coolant in the GCFR core are 50% coolant, 10% structural materials, and 40% fuel (which may include matrix material and cladding). The specific power (power per unit mass of fuel) is low due to the combination of low volumetric power density and relatively large fuel mass. Some typical numbers are: HTR 75 W/gHM, PWR 40 W/gHM, LMFBR 125 W/gHM, and GCFR 25-50 W/gHM [Duderstadt and Hamilton, 1976, Garnier et al., 2003]. To get an adequate density of heavy metal in the core, carbide or nitride fuels are preferred over oxide. For the present study nitride fuel was chosen, even though the economic feasibility of the required ^{15}N-enrichment is still doubtful [Inoue et al., 2002].

TRISO coated particle (CP) fuel has been used very successfully in thermal High Temperature Reactors (HTRs [Melese and Katz, 1984, Kugeler and Schulten, 1989]), and it is currently the reference fuel form for operating HTRs (HTTR [Lohnert, 2004], HTR-10 [Lohnert, 2002]) and HTRs under study (PBMR [Koster et al., 2003], NGNP [MacDonald, 2003], GT-MHR [Talamo and Gudowski, 2005], GTHTR300 [Kunitomi et al., 2004]). A redesign of the TRISO particle is necessary to adapt the HTR-type fuel particle for GCFR application. In figure 3.1 photos of TRISO particles are shown. Two redesigned coated fuel particles are proposed in this chapter: a small particle (diameter typically 1 mm) similar to TRISO CPs, and a hollow fuel sphere (HS: Hollow Sphere), an innovative design featuring a hollow shell of fuel surrounded by cladding, with a typical diameter of 3 cm.

Figure 3.1. *Left side: microscopic image of a TRISO particle, reproduced from Bailly et al. [1999]. Starting from the edge of the kernel are visible, the buffer layer, Inner Pyrolytic Carbon layer, the light colored SiC sealing layer and the Outer Pyrolytic Carbon layer. Right side: a bottle with HTTR TRISO particles. Photograph by the author.*

Coated particle fuel: description and GCFR adaption

A TRISO CP is made of a spherical fuel kernel (typical diameter 500 μm), surrounded by a buffer of porous graphite, a layer of pyrolytic carbon (IPyC, Inner Pyrolytic Carbon), a dense SiC sealing layer, and an outer layer of pyrolytic carbon (OPyC). The buffer provides voidage to accommodate kernel swelling and fission gas release during irradiation, and protects the cladding layers from recoiling fission fragments. The pressure within the particle can be modeled using the ideal gas law by:

$$p_{\text{buf}} = \frac{\text{FIMA} \cdot n_0 \cdot z \cdot k \cdot T_{\text{buf}}}{V_{\text{buf}}} \tag{3.1}$$

in which FIMA stands for *Fissions per Initial Metal Atom*, n_0 is the number of heavy metal atoms in the fuel kernel at Beginning of Cycle (BOC), z is the number of gas atoms released into the buffer per fissioned metal atom, k is Boltzmann's constant, and T_{buf} and V_{buf} are the temperature and void volume of the buffer layer, respectively. If the pressure within the particle is too large, the cladding layer will fail. A larger buffer volume will allow a higher burnup. The SiC layer is the main fission product release barrier and acts as a pressure vessel. The IPyC and OPyC layers contract under irradiation, thereby partly relieving the stress on the SiC layer induced by the increasing pressure of fission gases and kernel swelling during irradiation. The contraction rate of the pyrolytic carbon layers is roughly proportional to the average neutron energy [Martin, 2002a]. In the GCFR fast neutron spectrum, the contraction rate will probably be too large, and the layers will probably fail [Martin, 2002b].

A redesign of the TRISO particle is proposed: the IPyC and OPyC layers are removed, and the SiC sealing layer is replaced by a thin ZrC layer, which is more easily soluble than SiC, and chemically more stable at high temperatures [Pierson, 1996]. The Zr nuclei are more massive than Si nuclei, so damage (atom displacement) induced by collisions with high energy neutrons is expected to be less severe [Bailly et al., 1999]. In table 3.2 the GCFR CP is compared to modern HTR TRISO designs. The GCFR Coated Particle has a small buffer and relatively thin buffer and sealing layer, so the maximum burnup will be limited (several percent FIMA seems a reasonable estimate for lack of more detailed data). The temperature limits for the coated particles are the same as for HTR TRISOs, i.e. maximum operating temperature 1200°C and fission product retention up to 1600°C.

TRISOs can be combined with a matrix (e.g. graphite) to make fuel pebbles or compacts. The temperature gradients over such a fuel element can be considerable. See section 3.A for a detailed discussion of the temperature profile in a fuel pebble. For PBMR, the centerline temperature can be some 130 K higher than the gas temperature. The GCFR requires a larger power density than HTR (typically 10 - 20 times higher), easily leading to large temperature gradients within the fuel element.

The Random Close Packing (RCP) of spheres is roughly 63%, and the volume fraction of fuel inside a coated particle is $(r_k/r_t)^3$, with r_k the radius of the fuel kernel and r_t the radius of the TRISO particle. To make a fuel compact with coated particles satisfying the requirement of containing more than 45% of fuel by volume requires a coated particle with a very large kernel and almost no buffer and cladding. From these consideration it is clear that for coated particle

Table 3.2. *Geometry of contemporary TRISO designs. Values from Verfondern et al. [2001] for HTTR, Tang et al. [2002] for HTR-10, and Rütten and Kuijper [2003] for the HTR-N particle. The burnup targets for the particles are reflected by the ratio of kernel volume to buffer volume.*

Reactor	HTTR	HTR-10	HTR-N	GCFR
Design FIMA	3%	8%	80%	-
Kernel radius [μm]	300	249	120	350 - 380
Buffer thickness [μm]	60	95	95	100 - 70
IPyC thickness [μm]	30	42	40	n/a
SiC thickness [μm][a]	25	37	35	50
OPyC thickness [μm]	45	42	40	n/a
Total radius [μm]	460	465	330	500
Relative buffer volume $V_{\text{buf}}/V_{\text{kernel}}$	0.73	1.63	4.75	0.66 - 1.13

[a] ZrC for GCFR particle

fuel, direct cooling, i.e. a bed of particles with the coolant flowing in between the particles, is to be the most viable solution. An extra advantage is that the temperature differences between the fuel and the coolant remain small.

Hollow Sphere fuel description

The amount of voidage available in a coated particle is determined by the porosity of the buffer layer, which is usually about 50% for TRISO CPs. Removing all material in the buffer layer creates more empty space to accommodate kernel swelling and fission gas storage. This observation has led to the design of an innovative fuel particle: the hollow fuel sphere (HS: Hollow Sphere). The HS is a hollow shell of fuel material with ceramic cladding around it (figure 3.2). It is comparable to the fuel element proposed in [Ryu and Sekimoto, 2000] for GCFR applications.

Recoiling fission fragments will penetrate the cladding, but this should be no problem as long as the cladding thickness is much larger than the penetration depth of the fission fragments (typically several μm). An HS could be manufactured by pressing a mixture of fuel powder with a gelating agent to form hollow hemispheres. Two hemispheres are attached to each other and then sintered to form a full sphere, onto which a thick ceramic cladding layer is deposited. In a hollow fuel sphere the inner void is completely empty, providing more voidage than a TRISO CP with the same volume fraction of fuel and cladding.

3.4 Fuel subassembly design

In our design the fuel elements are cooled directly by helium. The coolant flow through the packed bed causes a pressure drop Δp_{pb}, which should not exceed some 2% of the system

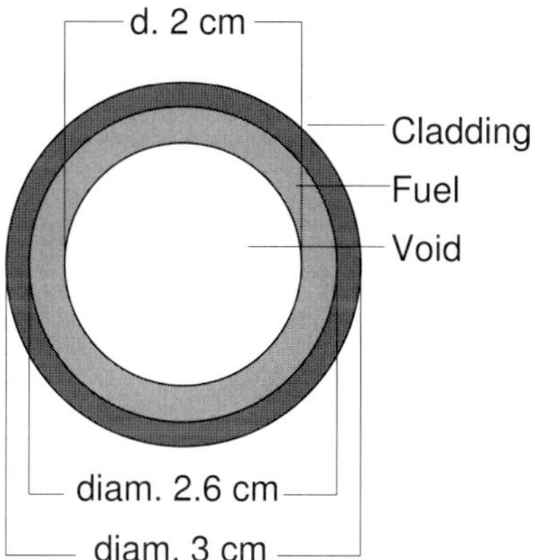

Figure 3.2. *A cross sectional view of the Hollow Sphere fuel element. The entire central void is available to accommodate fuel swelling and fission gas release.*

pressure in order to limit the required pumping power [Kugeler and Schulten, 1989]. It is necessary to find an expression to estimate the largest allowable height of the packed bed of fuel spheres. Δp_{pb} can be estimated using the Ergun-relation [Kinii and Levenspiel, 1991]:

$$\frac{\Delta p_{pb}}{h_{pb}} = 150\frac{(1-\epsilon)^2}{\epsilon^2}\frac{\mu_f}{d_p^2}\frac{u}{\epsilon} + 1.75\frac{1-\epsilon}{\epsilon}\frac{\rho_f}{d_p}\frac{u^2}{\epsilon^2} \tag{3.2}$$

in which h_{pb} is the bed height, ϵ is the porosity of the bed (37% for spheres in Random Close Packing), μ_f is the viscosity of the fluid, d_p is the diameter of the spherical particles, u is the superficial fluid velocity (i.e. the velocity the coolant would have if the particle bed was not present), and ρ_f is the density of the fluid. The superficial velocity is proportional to the mass flow rate $\dot{m}$:

$$\dot{m} = \rho_f A u \tag{3.3}$$

in which A is the geometrical cross-section of the outer dimensions of the pebble bed (i.e. A as if the pebbles weren't there. The packing factor ϵ is already taken into account in (3.2) in the u/ϵ-terms). The mass flow rate required to remove the heat from a packed bed of power producing spheres is proportional to the volume of the bed multiplied by the average power density of the bed P:

$$\dot{m} = \frac{P h_{pb} A}{c_p\left(T_{\text{core,out}} - T_{\text{core,in}}\right)} \tag{3.4}$$

with $T_{\text{core,in/out}}$ the coolant temperature at core inlet / outlet, and c_p the isobaric heat capacity of the fluid. It is assumed that the pressure drop over the bed is small compared to the system pressure, so that c_p can be taken as a constant. Combining the previous equations, and using

$T_{\text{core,out}} - T_{\text{core,in}} = \Delta T$ leads to a revised expression for the Ergun-relation:

$$\Delta p = 150\frac{(1-\epsilon)^2}{\epsilon^3}\frac{\mu_f}{\rho_f d_p^2}\frac{Ph_{pb}^2}{\Delta T c_p} + 1.75\frac{1-\epsilon}{\epsilon^3}\frac{1}{\rho_f d_p}\frac{P^2 h_{pb}^3}{\Delta T^2 c_p^2} \tag{3.5}$$

This alternative version of the Ergun-relation can be used to estimate the pressure drop over a bed of power producing spheres with a given volumetric power density using a given coolant at given temperature and pressure.

Table 3.3. *Design parameters for the Generation IV GCFR using coated particle fuel.*

Thermal power	2400 MWth
Power density P	50 MW/m^3
Core height h_c	3 m
Core radius r_c	2.25 m
$T_{\text{core,in}}$	450°C
$T_{\text{core,out}}$	850°C
Nominal pressure p	7 MPa
CP outside diameter	1 mm
HS outside diameter	3 cm

For a core with a volumetric power density $P = 50$ MW/m^3 (note that MW/m^3 is the same as W/cm^3, and all volumetric power densities in are taken as averaged over fuel and coolant), $h_{pb} = 7.5$ cm, $\epsilon = 0.37$, $T_{\text{core,in}} = 450$°C, $T_{\text{core,out}} = 850$°C and a particle diameter of 1 mm (Coated Particle fuel) the equation gives $\Delta p_{pb} = 0.04$ bar. For the Hollow Sphere fuel, the diameter of 3 cm and a bed height of 50 cm the equation gives $\Delta p_{pb} = 0.15$ bar. For a coated particle core, the height of the beds should thus not exceed several centimeters, for the HS core concept the bed height should not be larger than several tens of centimeters.

Coated particle fuel subassembly

A fuel subassembly design was made based on the core parameters listed in table 3.3. To obtain a low bed height for the CP fuel, a subassembly is proposed with two perforated concentric annuli, with the CPs sandwiched in between the perforated annuli (see figure 3.3). The height of the cylinder is h_c, while the particle bed thickness is around 2.5 cm. The overall diameter of such a fuel cylinder would be some 10 to 15 cm. The coolant enters the cylinder at the bottom, and exits at the top. The coolant flows inward to keep a compressive stress on the perforated cylinders. All parts are ceramics, e.g. the perforated tubes are made of SiC. The fuel beds occupy 75% of the cylinder volume. The fuel cylinders are arranged in a hexagonal lattice in the core.
The overall core volume fraction of coolant equals 57%, the volume fraction of fuel spheres is 43%, and the fuel volume fraction $(r_k/r_t)^3 * 0.43$. If annular hexagons are used instead of annular cylinders, the overall coolant volume fraction decreases and the fuel fraction may increase. The presented fuel cylinder concept is comparable to that presented in Chermanne

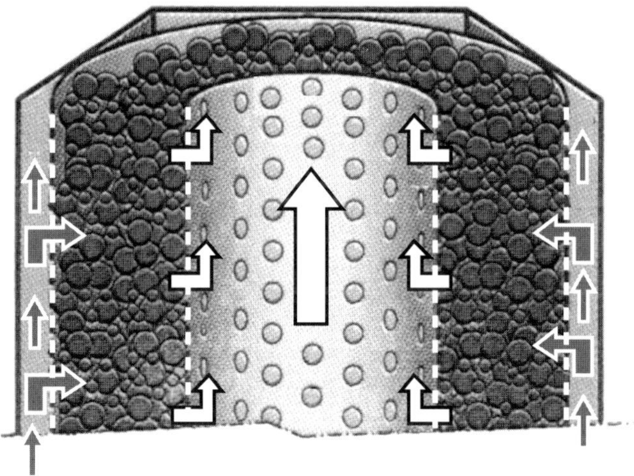

Figure 3.3. *Close up of an annular fuel cylinder, holding a bed of coated particles. The coolant enters at the bottom, flows inward through the packed bed of fuel particles and exits at the top. The inward flow assures a compressive stress on the inner cylinder. Dimensions: height 3 m, bed thickness ≈ 2.5 cm, overall diameter some 10 to 15 cm.*

[1972a], see section 2. A similar modern design is found in Konomura et al. [2003].

Hollow Fuel Sphere core layout

The HS concept allows a larger bed height, but the allowed bed height is still lower than the core height, so again a radial flow solution is chosen. The reactor core is divided in 3 concentric rings of fuel spheres, with the coolant entering from the top of the reactor, in between the beds. The coolant then flows radially through the beds, and exits the reactor. A schematic of this configuration is illustrated in figure 3.4. The height of the beds is the height of the reactor core, the thickness of the beds is about 50 cm. Restraining devices keep the fuel elements on their position in the core.

3.5 Burnup study of the coated particle GCFR

A burnup study was performed to estimate the performance of the coated particle GCFR in the closed fuel cycle. The fuel is made of ^{238}U and recycled LWR Pu. The Pu vector is taken from the HTR-N burnup benchmark [Rütten and Kuijper, 2003] and given by:

$$^{238}\text{Pu} / {}^{239}\text{Pu} / {}^{240}\text{Pu} / {}^{241}\text{Pu} / {}^{242}\text{Pu} = 1\% / 62\% / 24\% / 8\% / 5\%$$

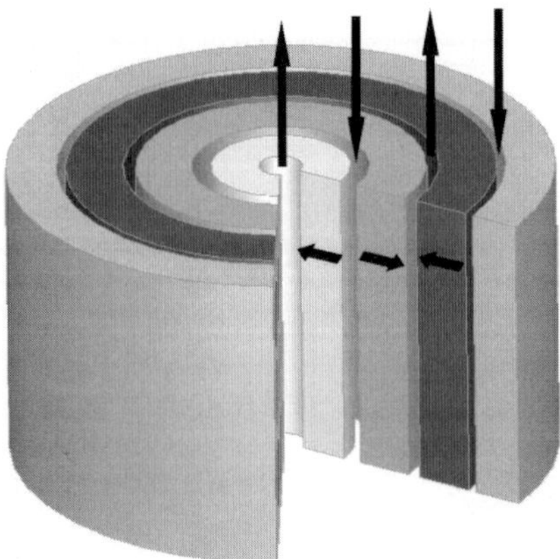

Figure 3.4. *The proposed core layout for the GCFR using Hollow Sphere fuel. The helium enters in between the beds, flows radially inward (or outward), then turns upward to leave the core. The outermost grey area is the reflector, the three inner annuli each contain a bed of Hollow Sphere fuel elements. The axial height is 3 m, the thickness of the fuel beds is some 50 cm.*

The specific power (W/gHM) is rather low for the GCFR, leading to long irradiation periods. The irradiation period is chosen as 1900 days (6 years at an average load factor of 85%, or 5.5 years at 95% load factor). This irradiation period results in a burnup of some 4-5% FIMA, depending on initial fuel loading. Pressure inside the coated particles due to fission gases should remain low for such a low burnup. The cool down period after irradiation is also 1900 days. Reprocessing is assumed to separate fission products from actinides. The actinides are then fully recycled, and depleted uranium is added to make the new fuel. For both the CP fuel element and the HS fuel element, two fuel compositions are calculated. The fuel composition differ in their plutonium content, and are designated as 'low plutonium fraction' and 'high plutonium fraction'. The fuel element geometry is changed according to the Pu-loading to obtain the same k_{eff} at BOC. The goal is to compare the performance in obtaining the closed fuel cycle, MA production, multi-cycle recycling etc.

There are 3 enrichment zones in the core, chosen to give a reasonably flat power density profile at startup. Simulations were done using various modules of the SCALE 4.4a code system [Oak, 2000], interconnected by a Perl script and some in-house Fortran codes. Cross sections were obtained from a JEF-2.2 172 group AMPX library [Hoogenboom and Kloosterman, 1997].

The calculation scheme is illustrated in figure 3.5. The calculations use the quasi-static approach: for each fuel material a cell calculation is done to obtain cell weighted cross sections, after which a whole core calculation is done to obtain k_{eff} and the flux profile. The flux profile is used to calculate the evolution of the fuel over a time interval, after which the process is repeated with the new nuclide densities. The quasi-static approximation is thus that the flux

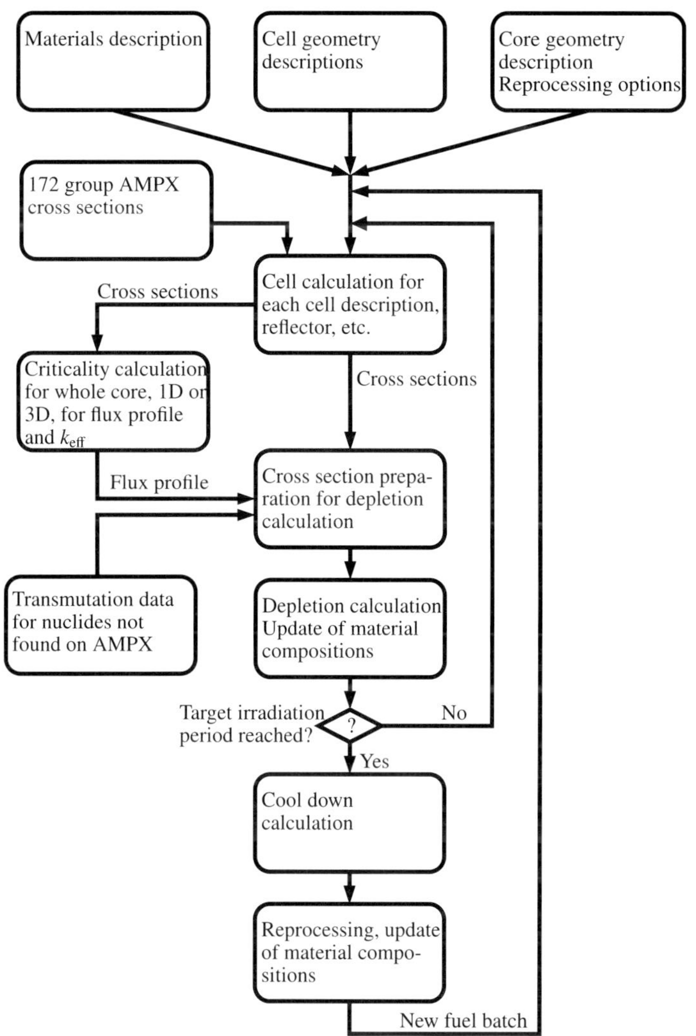

Figure 3.5. *The calculational path for an irradiation calculation. The cross sections obtained from the cell calculation are used to determine k_{eff} of the whole core, and the space and energy dependent flux profile in the core. This flux profile is then used to calculate the depletion of all materials in the reactor. This inner iteration is repeated until the target irradiation period is fully calculated. Then cool down and reprocessing are calculated. A new fuel description is prepared, and subsequently a new irradiation calculation is started.*

spectrum and the microscopic cross sections are held constant during a depletion step, after which a new spectrum is calculated for the next depletion step. When the final irradiation time is reached, cool down, reprocessing and the new fuel composition for the next batch are calculated. The Fuel Temperature Coefficient (FTC) was calculated from an extra criticality calculation at an elevated temperature $T_0 + \Delta T_0$. FTC is calculated using (see for instance de Kruijff and Janssen [1993]):

$$FTC = \frac{k_{\text{eff}}(T_0 + \Delta T_0) - k_{\text{eff}}(T_0)}{k_{\text{eff}}(T_0)} \cdot \frac{1}{\Delta T_0} \tag{3.6}$$

and expressed in pcm/K.

3.6 Results

A total of four sets of results were obtained from the burnup calculations (2 fuel concepts and 2 fuel specifications). Not all four sets will be analyzed here, but rather a summary of results is given using one set of results as a reference. The interested reader is referred to the article by Van Rooijen et al. [2005] in which a detailed analysis of all results is given.

In table 3.4 a summary is given of the fuel evolution during 4 batches of coated particle fuel (4 cycles of irradiation, cool down, integral reprocessing, and reloading into the reactor), and in table 3.5 the same data is given for the hollow fuel sphere concept. The initial plutonium fraction for all cases is between 12.5% and 15.5%. The lowest specific power is 20.45 W/gHM (Coated Particle core with low Pu fraction), the highest specific power is 27.86 W/gHM (Hollow Sphere core with high Pu fraction). The low plutonium cores have such a low fissile content that k_{eff} increases during irradiation due to (excessive) conversion of ^{238}U to ^{239}Pu. An example of this is given in figure 3.6 where k_{eff} and FTC are given for the hollow sphere core with low plutonium fraction. In section 3.9 the phenomenon of increasing reactivity during irradiation is treated in more detail.

Under the assumption of a 50% efficient power conversion system the GCFR will have 1200 MW electrical output. The low Pu cores contain 15.7 tonne Pu (CPs) and 14.3 tonne Pu (HS), so they are close to but within the tentative maximum of 15 tonne of Pu per GWe [Garnier et al., 2003]. The cores with high plutonium fraction contain a smaller *total amount* of Pu than the low Pu cores (about 13.5 tonne of Pu). The different HM loading (see table 3.4) is achieved for the CP core by using a smaller fuel kernel, keeping the total radius of the particle constant. For the HS core the thickness of the fuel shell is varied. A smaller fuel shell means a larger central void.

All cores have a negative FTC, and the magnitude of FTC decreases slightly during irradiation. The decrease is caused by the hardening of the spectrum due to fission product build-up during irradiation. In figure 3.7 the flux per unit lethargy is given at 0 days and 1900 days of irradiation, together with the resonances of σ_a of ^{238}U. After 1900 days of irradiation the flux is slightly decreased in the resonance region, leading to less resonance absorption and as a result a lower FTC. The coated particle core with high plutonium fraction, which uses a coated particle with more buffer volume, has the largest FTC, which is attributed to the extra moderation due to graphite in the buffer. Otherwise, the value of the FTC is comparable for all core concepts (about -2 pcm/K).

Table 3.4. *Results of the burnup study using coated particle fuel (integral recycling, adding only ^{238}U).*

	Low plutonium fraction	*High plutonium fraction*
	Batch 1	
HM total [kg]	117379	91725
– of which Pu [kg]	15733 / 13.4%	13830 / 15.1%
^{238}Pu / ^{239}Pu	1% / 62%	1% / 62%
^{240}Pu / ^{241}Pu / ^{242}Pu	24% / 8% / 5%	24% / 8% / 5%
ΔPu / $\Delta Pu_{fissile}$	+2.2% / -0.7%	-0.7% / -6.1%
EOC MA mass [kg]	514	592
k_{max} / k_{min}	1.0651 / 1.0491	1.0658 / 1.0116
FTC_{max} / FTC_{min} [pcm/K]	-2.07 / -1.42	-2.6 / -1.5
	Batch 2	
HM total [kg]	117387	91731
– of which Pu [kg]	16083 / 13.8%	13728 / 15 %
^{238}Pu / ^{239}Pu	0.85% / 63.7%	0.87% / 61.7%
^{240}Pu / ^{241}Pu / ^{242}Pu	26.3% / 4.3% / 4.9%	27.9% / 4.4% / 5.1%
ΔPu / $\Delta Pu_{fissile}$	+3.7% / +1.3%	+2.0% / -2.1%
EOC MA mass [kg]	734	850
k_{max} / k_{min}	1.0449 / 1.0333	1.0045 / 0.9941
FTC_{max} / FTC_{min} [pcm/K]	-1.79 / -1.48	-2.5 / -1.3
	Batch 3	
HM total [kg]	117392	91736
– of which Pu [kg]	16672 / 14.3%	14002 / 15.4 %
^{238}Pu / ^{239}Pu	1.1% / 63.4%	1.2% / 60.0%
^{240}Pu / ^{241}Pu / ^{242}Pu	27.9% / 3.1% / 4.6%	30.4% / 3.5% / 4.8%
ΔPu / $\Delta Pu_{fissile}$	+3.3% / +1.3%	+2.3% / -0.7%
EOC MA mass [kg]	851	994
k_{max} / k_{min}	1.0521 / 1.0360	0.9910 / 0.9959
FTC_{max} / FTC_{min} [pcm/K]	-1.88 / -1.33	-3.0 / -0.96
	Batch 4	
HM total [kg]	117397	91740
– of which Pu [kg]	17230 / 14.8%	14329 / 15.7%
^{238}Pu / ^{239}Pu	1.4% / 62.4%	1.5% / 58.3%
^{240}Pu / ^{241}Pu / ^{242}Pu	29.2% / 2.7% / 4.3%	32.2% / 3.4% / 4.6%
ΔPu / $\Delta Pu_{fissile}$	+2.7% / +1.0%	+2.1% / -0.2%
EOC MA mass [kg]	930	1096
k_{max} / k_{min}	1.0594 / 1.04396	0.9891 / 0.9972
FTC_{max} / FTC_{min} [pcm/K]	-1.77 / -1.33	-2.2 / -1.3
	Geometry of the fuel element	
Kernel radius r_k [μm]	380	350
Buffer radius r_b [μm]	450	450
Cladding radius r_c [μm]	500	500

Table 3.5. *Results of the burnup study using hollow fuel spheres (integral recycling, adding only ^{238}U).*

	Low plutonium fraction	*High plutonium fraction*
	Batch 1	
HM total [kg]	114585	86144
– of which Pu [kg]	14336 / 12.5%	13374 / 15.5%
^{238}Pu / ^{239}Pu	1% / 62%	1% / 62%
^{240}Pu / ^{241}Pu / ^{242}Pu	24% / 8% / 5%	24% / 8% / 5%
ΔPu / $\Delta Pu_{fissile}$	+4.4% / +2.7%	-1.8% / -6.3%
EOC MA mass [kg]	458	427
k_{max} / k_{min}	1.0294 / 1.0331	1.0844 / 1.0307
FTC_{max} / FTC_{min} [pcm/K]	-2.3 / -1.2	-1.84 / -1.41
	Batch 2	
HM total [kg]	114552	86148
– of which Pu [kg]	14960 / 13.1%	13131 / 15.3 %
^{238}Pu / ^{239}Pu	0.8% / 64.8%	0.9% / 62.5%
^{240}Pu / ^{241}Pu / ^{242}Pu	25.5% / 4.1% / 4.8%	27.3% / 4.3% / 5.1%
ΔPu / $\Delta Pu_{fissile}$	+4.8% / +3.2%	+1.1% / -2.2%
EOC MA mass [kg]	645	600
k_{max} / k_{min}	1.0396 / 1.0174	1.0233 / 1.0128
FTC_{max} / FTC_{min} [pcm/K]	-2.1 / -1.2	-1.76 / -1.23
	Batch 3	
HM total [kg]	114557	86151
– of which Pu [kg]	15681 / 13.8%	13281 / 15.5 %
^{238}Pu / ^{239}Pu	1% / 65%	1.2% / 61.3%
^{240}Pu / ^{241}Pu / ^{242}Pu	26.7% / 2.8% / 4.4%	29.4% / 3.2% / 4.8%
ΔPu / $\Delta Pu_{fissile}$	+3.9% / +2.4%	+1.6% / -0.7%
EOC MA mass [kg]	742	688
k_{max} / k_{min}	1.0516 / 1.0310	1.0141 / 1.0100
FTC_{max} / FTC_{min} [pcm/K]	-2.0 / -0.7	-1.73 / -1.48
	Batch 4	
HM total [kg]	114562	86154
– of which Pu [kg]	16290 / 14.3%	13500 / 15.8%
^{238}Pu / ^{239}Pu	1.3% / 64.3%	1.5% / 60%
^{240}Pu / ^{241}Pu / ^{242}Pu	27.8% / 2.5% / 4.1%	30.9% / 3.0% / 4.6%
ΔPu / $\Delta Pu_{fissile}$	+3.0% / +1.5%	+1.5% / -0.3%
EOC MA mass [kg]	806	746
k_{max} / k_{min}	1.0625 / 1.0451	1.0153 / 1.0087
FTC_{max} / FTC_{min} [pcm/K]	-1.8 / -0.8	-1.88 / -1.28
	Geometry of the fuel element	
Sphere diameter [cm]	3	3
Cladding thickness [cm]	0.2	0.2
Fuel thickness [cm]	0.325	0.225

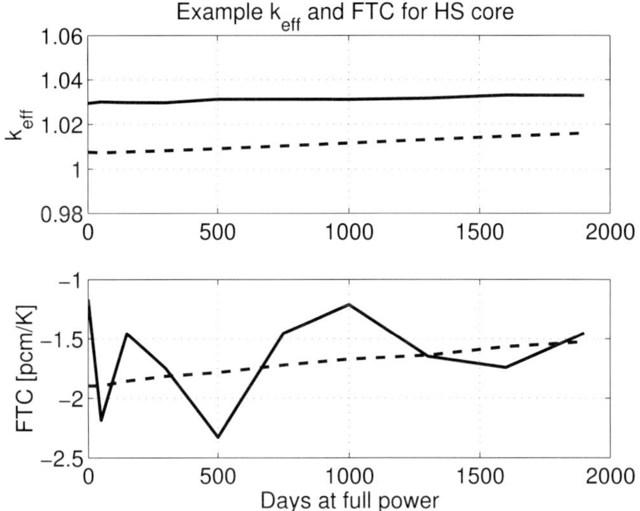

Figure 3.6. *Typical time dependence of k_{eff} in the hollow sphere GCFR core. The positive reactivity swing is a result of (excessive) production of ^{239}Pu from ^{238}U. In this graph, results for k_{eff} and FTC from a 1D (dashed line) and a 3D (solid line) calculation are shown. In the 3D calculation the axial reflectors are treated properly, resulting in a consistently higher k_{eff}. However, the trend of the 1D and 3D graphs are similar. The shape of the 3D FTC result is somewhat chaotic. This is due to the statistical nature of the Monte-Carlo calculations with KENO-V. However, the magnitude of the FTC for 1-D and 3-D calculations is very similar, as is the general trend (decrease) with burnup.*

For the coated particle core the pressure within the particles was evaluated using equation (3.1). At the end of an irradiation cycle, the burnup reaches some 4% FIMA. For z, the number of gas atoms released into the buffer per HM atom fissioned, a value of 0.8 was adopted, which is a very conservative value. For thermal HTRs, $z = 0.3 - 0.4$ is commonly used, but accurate values are not known for the GCFR non-oxide fuel. Using $z = 0.8$, the pressure at the end of the burnup interval is estimated between 2 and 4 MPa, so the particles are always under compressive stress during irradiation.

In all cores there is a net production of plutonium. The production is larger in the low plutonium cores. On the other hand, the production of MA is higher in the high plutonium cores. This is to be expected, as plutonium is a source term for higher actinides, i.e. Am and Cm. In the course of the four batches, the Pu fraction increases, and hence the production of plutonium decreases. Extrapolating from the tables, a 'break even' core would have a Pu fraction of around 16%. The Pu vector becomes quite degradated over the irradiation cycles, with a marked decrease of ^{241}Pu. Also note that although the amount of MA in the core at BOC steadily increases, the total amount of minor actinides in the core remains limited to about 1% of the fuel mass.

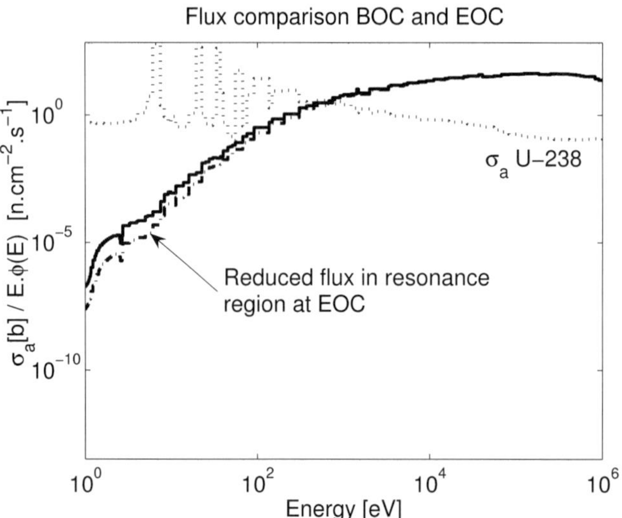

Figure 3.7. *The flux per unit lethargy as a function of energy after 0 and 1900 days of irradiation in the HS core. The resonances of σ_a of ^{238}U are also shown. After 1900 days the spectrum has hardened, giving a lower flux in the ^{238}U resonances, leading to lower absorption and a decrease in the magnitude of the FTC. The BOC spectrum of the CP core is indicated by dots. This spectrum is softer than the HS spectrum.*

3.7 Extended time calculations

To examine the long term evolution of the fuel a calculation was done over 8 fuel batches. One result is given here, namely the result for the Hollow Sphere core with high plutonium fraction. This core was chosen because it has the lowest fuel mass and lowest plutonium fraction of all cores (thus highest specific power, and best fuel economy), and has *BG* closest to zero of all cores. MA production is lowest in this core. The other core configurations studied show a somewhat worse performance on all these points. In table 3.6 a short summary of the results is given. Also given is the EOC MA mass for each fuel batch.

In figure 3.8 the evolution of plutonium isotopes during irradiation is given as an illustration. There is almost no change in the amount of fissile nuclides in the fuel, which is a requirement to obtain a closed fuel cycle, without blankets and without breeding extra fissile material. In that sense, the Generation IV objective of the closed fuel cycle is within reach using the Hollow Fuel Sphere GCFR concept.

From table 3.6 it is seen that in fact an equilibrium is almost reached after 8 fuel batches. The plutonium production becomes almost zero, the amount of MA becomes constant (MA production in equilibrium with MA destruction). The FTC is always negative. The magnitude of FTC is influenced by the presence of MA in the fuel. The extra absorption introduced by some MA, most notably ^{237}Np and ^{241}Am, decreases the flux in the resonance energy range,

Table 3.6. *Results of the burnup study over 8 batches for the Hollow Sphere core with high Pu fraction.*

	Hollow Sphere fuel core High plutonium fraction
Batch 1	
Pu initial	15.5%
ΔPu / $\Delta Pu_{fissile}$	-1.8% / -6.3%
EOC MA mass [kg]	427
Batch 2	
Pu initial	15.3 %
ΔPu / $\Delta Pu_{fissile}$	+1.1% / -2.2%
EOC MA mass [kg]	600
Batch 3	
Pu initial	15.5 %
ΔPu / $\Delta Pu_{fissile}$	+1.6% / -0.7%
EOC MA mass [kg]	688
Batch 4	
Pu initial	15.8%
ΔPu / $\Delta Pu_{fissile}$	+1.5% / -0.3%
EOC MA mass [kg]	746
Batch 5	
Pu initial	16.0 %
ΔPu / $\Delta Pu_{fissile}$	+1.3% / -0.1%
EOC MA mass [kg]	790
Batch 6	
Pu initial	16.2 %
ΔPu / $\Delta Pu_{fissile}$	+1.0% / -0.1%
EOC MA mass [kg]	824
Batch 7	
Pu initial	16.4 %
ΔPu / $\Delta Pu_{fissile}$	+0.8% / -0.1%
EOC MA mass [kg]	851
Batch 8	
Pu initial	16.6 %
ΔPu / $\Delta Pu_{fissile}$	+0.7% / -0.0%
EOC MA mass [kg]	872

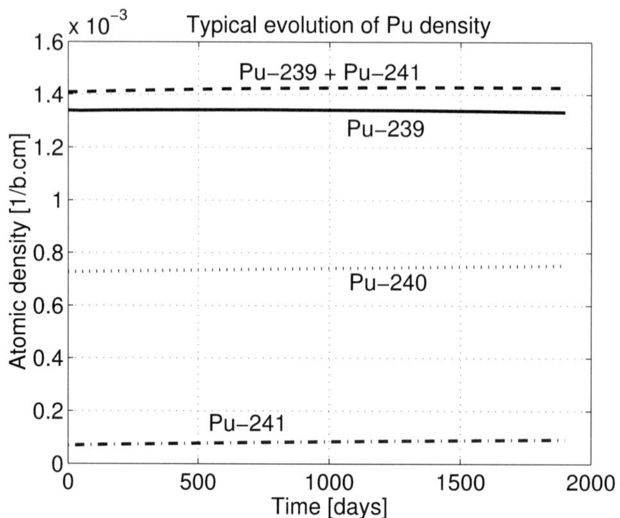

Figure 3.8. *Typical evolution of Pu isotopes in the hollow sphere GCFR. During irradiation, the amount of fissile material remains almost constant, which is a prerequisite to obtain a closed fuel cycle.*

but since the MA loading remains limited to around 1%, the effect is not very large. FTC is between -1.5 and -2 pcm/K for all calculations.

3.8 Moving on to GFR600

The coated particle GCFRs show excellent performance regarding the closed fuel cycle, and the study on the coated particle GCFR was very successful. When the European FP6 GCFR-STREP[1], started in January 2005, the partners agreed to adopt the CEA[2] GFR600 GCFR design as the reference. GFR600 has plate fuel rather than coated particle fuel. The main reasons to not further pursue the coated particle design were:

- Given the expected availability of plutonium, the GCFR specific power should be at least 50 MW/gHM to enable deployment of a reasonable number of GCFR plants with a closed fuel cycle. The target for volumetric power density is set at 100 MW/m^3.
- The coated particle cores with their radial coolant flow have a very doubtful performance for natural convection under accidental conditions.

[1] European Commission / Euratom 6th Framework Programme 'Gas Cooled Fast Reactor Specific Targeted REsearch Programme' (GCFR-STREP)

[2] Commissariat à l'Energie Atomique, the French research organization for nuclear power

- The TRISO particles are in direct contact with the coolant, with only one barrier (the coating layer(s)) for fission product release into the primary system. In Generation IV it is preferred to have several barriers for FP release, such as a matrix fuel: fission products are stored within the matrix, which itself is encased in cladding.

GFR600 has a volumetric power density of 100 MW/m^3, a specific power of 45 W/gHM, and uses a plate type fuel. The fuel is a matrix fuel, with 70% UPuC and 30% SiC by volume. The plates are clad with SiC, and the reflector material is Zr_3Si_2. The unit power of 600 MWth was chosen to enable a 'modular' system. An illustration of a fuel assembly and the core layout of GFR600 is given in figure 3.9. More design data of GFR600 are given in table 5.1. The application of SiC for cladding and structural elements is very challenging. Limited experience exists with SiC as a nuclear material (cladding layer of TRISO particles, and see Every [2006] for examples of SiC as a cladding for pin fuel). Within the GCFR STREP, CEA is confident that the proposed fuel element can be manufactured, and the layout of figure 3.9 was adopted as the reference design.

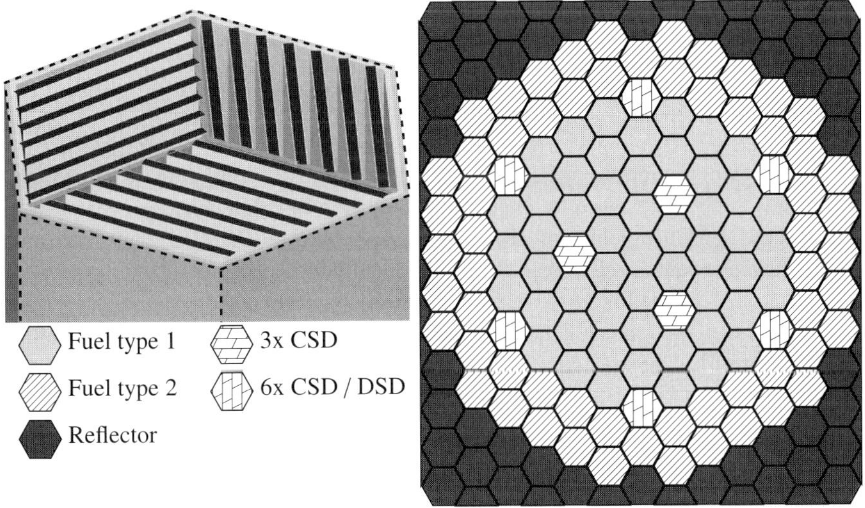

Figure 3.9. *A GFR600 plate fuel assembly, and the layout of the core. CSD = Control Shut Down assembly, DSD = Diverse Shut Down assembly. The core is composed of 112 fuel assemblies and is surrounded by 210 reflector assemblies.*

For GFR600 burnup calculations were performed focusing on transmutation of Minor Actinides (Np, Am and Cm), homogeneously mixed into the fuel. The isotopic vector of MA to be added to the fuel is given in table 3.7. This material is representative of the MA available in the 21st century from LWRs. This vector is based on scenario studies, employing a PWR reactor fleet with a partial MOx fuel cycle. These scenario studies were performed by CEA and the vector shown in table 3.7 is chosen as the reference in the STREP. Calculations

Table 3.7. *The isotopic composition of the Minor Actinides added to the fuel in the GFR600 MA transmutation study. The bulk of this material is ^{237}Np and ^{241}Am, which have a large absorption cross section, and transmute to fissile materials (^{237}Np to ^{238}Pu and ^{241}Am to ^{242}Am, see also chapter 5).*

			Cumulative
Np	16.86%		
^{237}Np		100%	16.86 %
Am	76.56%		
^{241}Am		79.2%	60.64 %
^{242m}Am		0.3%	0.23 %
^{243}Am		20.5%	15.69 %
Cm	6.59%		
^{242}Cm		0.3%	0.02 %
^{243}Cm		1.1%	0.07 %
^{244}Cm		78.0%	5.14 %
^{245}Cm		19.1%	1.25 %
^{246}Cm		1.5%	0.10 %
Total			100%

were done substituting up to 10% of the uranium with this MA mixture. The rationale of this choice is that it was shown by the multiple recycling calculations on the coated particle GCFR that the MA fraction in the fuel cycle remains limited to about 1%. Thus there may be room to transmute some extra MA in the GCFR core, in order to reduce existing stock-piles. The absorption cross section of the actinides, especially ^{237}Np and ^{241}Am, is larger than σ_a of ^{238}U which they are replacing. This increases the total absorption in the core, resulting in a harder spectrum, as illustrated in figure 3.10. The spectrum of GFR600 with MA is harder than that of GFR600 using the UPuC fuel, but it is softer than the spectrum of the large, oxide fueled, sodium cooled fast reactor CAPRA [Rouault et al., 1994].

For GFR600, and more generally for any homogeneous fast reactor running in a closed fuel cycle, the burnup reactivity swing may be positive under certain conditions (see section 3.9). For GFR600, the burnup reactivity swing becomes smaller with increasing MA loading, and becomes positive if more than 5% of MA are added to the fuel. This is caused by the MA: many of them are strong absorbers, reducing k_{eff} at BOC. During irradiation, the MA transmute to fissile material, resulting in a higher k_{eff}. The positive reactivity swing requires some external reactivity control, but this is not expected to be problematic because of long time scales involved. Figure 3.11 gives the evolution of k_{eff} with time of GFR600 as a function of the MA loading of the fuel.

From figure 3.11 it might be concluded that the reference fuel, without extra MA, results in a high burnup reactivity loss, but this is not true. For the reference fuel, the calculation during 1300 days results in a burnup of about 5% FIMA. Over that period, only about 6$ of reactivity is lost. This is much less than for instance a PWR. A calculation was done for EPR using

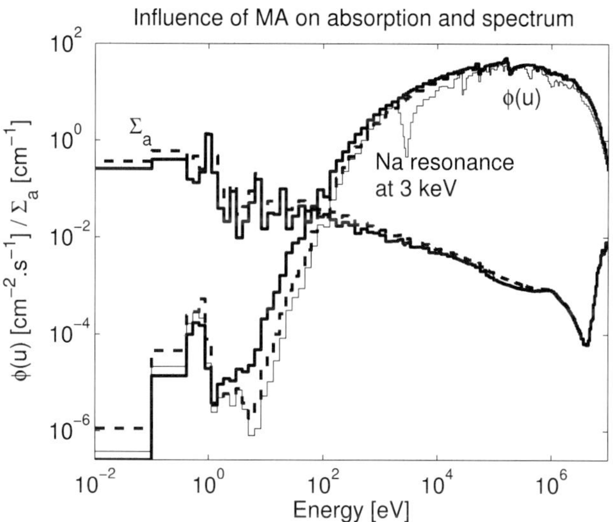

Figure 3.10. *Macroscopic absorption cross section Σ_a, and the neutron spectrum of GFR600. The solid lines are for the reference UPuC fuel, and the dashed lines for a fuel with 5% MA admixed (cell calculations). The fine line represents the spectrum of the CAPRA reactor, a large sodium-cooled, $UPuO_2$-fueled fast reactor. The reference GFR600 spectrum is softer than CAPRA. Adding MA gives more absorption at lower energies, resulting in a harder spectrum.*

MOx fuel as described in Leppanen [2005]. This calculation shows that the reactivity swing is some 24$ for MOx fuel, irradiated to 4% FIMA in 36 months. For GFR600, the fuel with 5% MA added gives a very low reactivity swing of about 4$ between BOC and 10% FIMA. Using an adequate fuel specification with some MA admixed, very long irradiation intervals can be achieved for GFR600 without the need of compensating a very large over-reactivity at BOC.

3.9 Positive reactivity swing and the performance parameter h

The coated particle GCFRs and GFR600 show that the evolution of k_{eff} during irradiation depends strongly on the fuel composition, and a positive reactivity swing might occur. This cannot be solely due to the requirement of $BG = 0$, i.e. no net fissile consumption, because fission products introduce extra absorption, and slightly absorbing HM may become a strong absorber by transmutation. Also, it is not solely due to the homogeneous fuel, because criticality of the reactor is determined by both the isotopic composition and spatial distribution of the fuel. Following Permana et al. [2006], consider the performance parameter h, which is

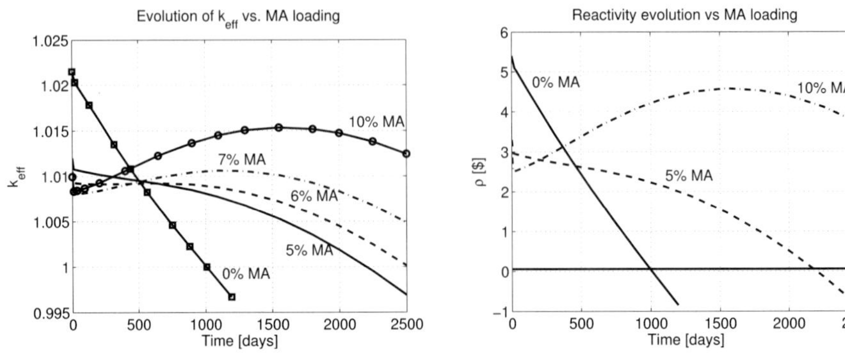

Figure 3.11. *Left: Evolution of k_{eff} as a function of time for various MA loadings. From about 6% MA, the burnup reactivity swing becomes positive. For GFR600, the irradiation interval of 2500 days results in a burnup of roughly 10% FIMA. For comparison, the evolution of k_{eff} for the reference fuel without MA is also given. Right: a subset of the data from the left graph, with the reactivity now expressed in \$. The reference (0% MA) fuel has a reactivity loss of some 6\$ over 1300 days (= 5% FIMA). The fuel with 5% MA can reaches 10% FIMA after 2500 days, with a reactivity swing of only 4\$.*

defined by:

$$h \equiv \frac{\langle \vec{N}, \nu\vec{\sigma}_f \rangle}{\langle \vec{N}, \vec{\sigma}_a \rangle} \tag{3.7}$$

with $\vec{N}$ the vector of nuclide densities, $\nu\vec{\sigma}_f$ the neutron production from fissions, and $\vec{\sigma}_a$ the vector with absorption cross sections per nuclide. The brackets $\langle , \rangle$ denote scalar products, i.e. integration over space, time and energy, and summation over all nuclides in the system under consideration. This parameter h is related to the multiplication factor of an infinite homogeneous nuclear reactor in a 1-group formalism:

$$k_\infty = \frac{\nu\Sigma_f}{\Sigma_a} \tag{3.8}$$

A GCFR with a homogeneous core (i.e. no blankets) is quite well approximated in the homogeneous 1-group formalism, and then $h = k_\infty$. The effective multiplication factor k_{eff} equals k_∞ times a leakage factor:

$$k_{\text{eff}} = \frac{\nu\Sigma_f}{\Sigma_a}\frac{1}{1 + L_D^2 B_g^2} = \frac{h}{1 + L_D^2 B_g^2} \tag{3.9}$$

In the closed fuel cycle, only fertile material is added to the recycled material to make new fuel. This means that only the $\langle \vec{N}, \vec{\sigma}_a \rangle$-term of equation (3.7) can be adjusted by the user during reprocessing (fertile material has a negligible fission cross section). It seems reasonable

to assume that to obtain a good performance of the reactor (irradiation interval etc.), the term $\langle \vec{N}, \nu\vec{\sigma}_f \rangle$ should roughly be the same at BOC for each new irradiation cycle. In the closed fuel cycle, the fission term cannot be influenced directly, as only fertile material is to be added to the reprocessed mixture.Thus the $\langle \vec{N}, \nu\vec{\sigma}_f \rangle$ term should be roughly preserved during the irradiation and subsequent cool down to obtain a closed fuel cycle (cool down period: the fuel is stored on-site for some time to allow the radioactive fission products to decay).
During irradiation, absorption is decreased by the consumption of heavy nuclides, and increased by the generation of fission products. If highly absorbing nuclides are present in the mixture, the $\langle \vec{N}, \vec{\sigma}_a \rangle$ can decrease quickly due to transmutation of the absorbers. Thus in a homogeneous reactor, designed to maintain the fission term $\langle \vec{N}, \nu\vec{\sigma}_f \rangle$ during irradiation, the parameter h can easily increase during irradiation if strong absorbers are present at BOC. This effect is even stronger if absorbing nuclides transmute to fissile ones (for example, ^{241}Am (absorber) becoming ^{242}Am (fissile)). For the GCFR, k_{eff} is approximated by $h/(1 + L_D^2 B_g^2)$, hence, a positive reactivity swing may occur if absorbers are present in the fresh fuel, especially if these absorbers are transmuted to fissile isotopes. If a positive reactivity swing is unwanted, a heterogeneous core design can be used where fissile material is bred in a location of the reactor where it does not contribute significantly to the multiplication k_{eff}.
Applying a quasi-static approximation where the cross sections σ_f, σ_a in (3.7) are taken as constants, the time derivative of h is:

$$\frac{dh}{dt} = \frac{1}{(\langle \vec{N}, \vec{\sigma}_a \rangle)^2} \{ \langle \frac{d\vec{N}}{dt}, \nu\vec{\sigma}_f \rangle \langle \vec{N}, \vec{\sigma}_a \rangle - \langle \frac{d\vec{N}}{dt}, \vec{\sigma}_a \rangle \langle \vec{N}, \nu\vec{\sigma}_f \rangle \} \tag{3.10}$$

The terms involving $\vec{N}$ are always positive, so the sign of dh/dt is determined by the terms involving $d\vec{N}/dt$. In a thermal reactor, the sign of $\langle \frac{d\vec{N}}{dt}, \nu\vec{\sigma}_f \rangle$ is always negative, because the production of fissile material per fission is low. The magnitude of the $\langle \frac{d\vec{N}}{dt}, \vec{\sigma}_a \rangle$ is usually small (decrease of absorption due to HM is offset by increase of absorption due to fission products). In a fast reactor with in-core breeding, and a closed fuel cycle, the $\langle \frac{d\vec{N}}{dt}, \nu\vec{\sigma}_f \rangle$-term is roughly zero, resulting in the possibility of a positive reactivity swing if the decrease in absorption due to transmutation of absorbers outweighs the increase due to production of fission products.

3.10 Conclusions

In this chapter two GCFR concepts are investigated, namely a GCFR concept using Coated Particle fuel, and a concept using plate fuel. The investigations of the coated particle GCFR concept yield the following conclusions:

- The coated particle cores (TRISO-like coated particles and Hollow Sphere fuel) presented in this chapter are able to obtain a closed fuel cycle, in a 2400 MWth reactor, with 50 MW/m^3 power density.

- In the closed fuel cycle the MA content of the fuel remains limited to between 1% and

1.5% of the total fuel mass, as illustrated in tables 3.4 and 3.5.

- The TRISO coated particle fuel as used in thermal HTRs is not applicable for GCFRs. Rather, a re-designed particle is necessary. The redesigned particle should have a large fuel fraction to allow a critical reactor with a plutonium fraction low enough to meet the target of self-breeding.

- The coated particle concept, whether the coated particles or the hollow fuel sphere concept is concerned, yields a low overall volume fraction of fuel in the core. To meet the demand of self-breeding, the fissile fraction of the fuel should be rather low. Criticality is problematic with such a low enrichment, leading to the preference of using a large core to reduce neutron losses by leakage, i.e. large unit power with medium power density. Note: a large power density could also be chosen, but then the fraction of fuel may have to be reduced to not exceed pumping power limits. Then the benefit of large core size is (partly) lost.

- The application of compound fuel elements, such as fuel pebbles or compacts, should be avoided for several reasons: firstly, the fuel volume fraction in such fuel elements would be very low, secondly, the temperature gradients inside the fuel elements would be very large (see section 3.A). Finally, the pressure drop over a pebble bed would be too large if a reasonable power density is chosen. Therefore, coated particles in GCFRs should be cooled directly by the coolant, without encapsulation in larger fuel elements.

- Although the coated particle cores presented in this chapter have a relatively large fraction of coolant, and a low fraction of fuel in the core, they have a low specific power between 21 and 28 W/gHM, see section 3.6. This value is lower than the target values for Generation IV (50 W/gHM). The low specific power is a drawback of this GCFR concept.

Concerning the plate-type GFR600 for which MA transmutation was investigated, the following conclusions are drawn:

- In the closed fuel cycle, the MA loading of the fuel remains limited to some 1% to 2%. Extra MA can be added to the fuel, if the fuel density is slightly increased to obtain criticality. If more than 6% of MA material is added to the fuel, the burnup reactivity swing becomes positive.

- Any reactor with homogeneous fuel, designed to run in a closed fuel cycle, will have a low burnup reactivity swing in general. For certain fuel compositions, the burnup reactivity swing may be positive. With a proper fuel design, incorporating some 5% MA, long irradiation intervals may be obtained with low reactivity swing ($\approx$ 4$), reducing the need to compensate a large overreactivity at BOC.

3.A Appendix: Temperature profile in a fuel pebble

A fuel pebble as used in HTRs consists of a central fuel zone (zone I with radius r_{fz}), surrounded by a layer of pure graphite (zone II, radius r_{peb}). The fuel zone itself is made of a mixture of TRISO fuel particles embedded in the graphite matrix. The temperature profile within the pebble is a function of the power produced within the pebble, and the material properties of the fuel zone, outer graphite layer and the coolant. To determine the temperature profile inside a fuel pebble, it is assumed for the moment that the TRISO-particles are evenly dispersed, so that the inner fuel zone of the pebble has a uniform volumetric power density. For the temperature in the three spatial domains of the pebble (fuel zone, graphite shell and coolant) the following equations apply [Kugeler and Schulten, 1989]:

$$\frac{1}{r^2}\frac{d}{dr}\left(r^2\lambda_I\frac{dT_I}{dr}\right)+q=0 \quad \text{[fuel zone]} \tag{3.11}$$

$$\frac{1}{r^2}\frac{d}{dr}\left(r^2\lambda_{II}\frac{dT_{II}}{dr}\right)=0 \quad \text{[graphite shell]} \tag{3.12}$$

$$4\pi r_{peb}^2 h\left(T_L-T_G\right)-q\frac{4\pi}{3}r_{fz}^3=0 \quad \text{[pebble to coolant]} \tag{3.13}$$

Here, q is the volumetric power density in the fuel zone, assumed to be uniform, and $\lambda_{I/II}$ is the thermal conductivity of the material in zone I/II. h is the heat transfer coefficient for heat transfer from the pebble surface to the coolant, and $T_L - T_G$ is the temperature difference between the pebble surface and bulk coolant. q is related to the volumetric core power density P by:

$$q=\frac{P}{1-\epsilon}\left(\frac{r_{peb}}{r_{fz}}\right)^3 \tag{3.14}$$

where ϵ is the porosity of the pebble bed ($\approx 37\%$). The system of differential equations has boundary conditions:

$$\begin{aligned}
T_I &= T_{II} && \text{for} \quad r=r_{fz}\\
\frac{dT_I}{dr} &= \frac{dT_{II}}{dr} && \text{for} \quad r=r_{fz}\\
T_{II} &= T_L && \text{for} \quad r=r_{peb}\\
\frac{dT_I}{dr} &= 0 && \text{for} \quad r=0\\
T_I &= T_m && \text{for} \quad r=0
\end{aligned} \tag{3.15}$$

where T_m indicates the maximum temperature. The solutions to equations (3.11) and (3.12) are:

$$T_I(r) = \frac{-qr^2}{6\lambda} + T_m \tag{3.16}$$

and

$$T_{II}(r) = \frac{2qr_{fz}^3}{6\lambda r} - \frac{qr_{fz}^2}{2\lambda} + T_m \tag{3.17}$$

These solutions are shown in figure 3.12. Solving for the difference over the three domains results in:

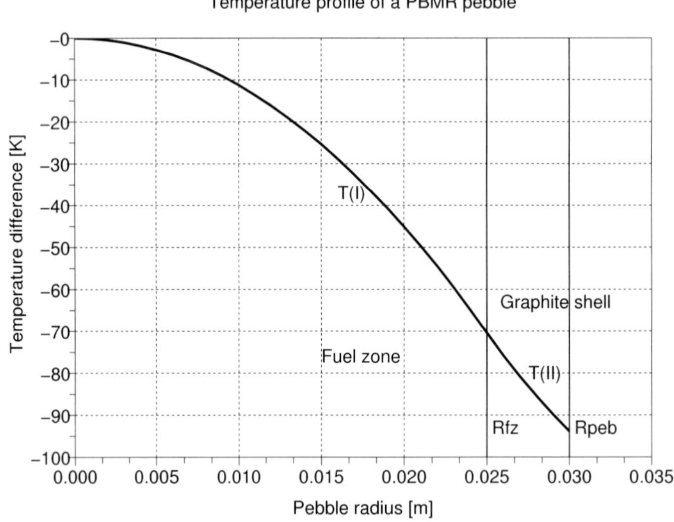

Figure 3.12. *Temperature profile in an HTR fuel pebble, from equations* (3.16) *and* (3.17). *All parameters are as in the illustrative example in the text.*

$$\begin{aligned}
\Delta T_I &= \frac{qr_{fz}^2}{6\lambda} = \frac{q_p}{8\lambda\pi r_{fz}} \\
\Delta T_{II} &= \frac{qr_{fz}^2}{3\lambda}\left(1 - \frac{r_{fz}}{r_{peb}}\right) = \frac{q_p}{4\pi\lambda}\left(\frac{1}{r_{fz}} - \frac{1}{r_{peb}}\right) \\
\Delta T_C &= \frac{4\pi qr_{fz}^3}{3hA} = \frac{q_p}{4\pi h}\frac{1}{r_{peb}^2}
\end{aligned} \tag{3.18}$$

where $q_p = q\frac{4}{3}\pi r_{fz}^3$ is the power produced in the pebble. Taking some typical values from the NEA PBMR benchmark [Nuclear Energy Agency, 2005]: r_{fz} = 2.5 cm, r_{peb} = 3 cm, h = 2665 W/m^{2}K (typical for HTRs like PBMR), λ = 20 W/mK, and q_p = 885 W (corresponding to an average core power density of 4.94 MW/m^3), it is found that ΔT_I = 70.43 K, ΔT_{II} = 23.48 K, ΔT_C = 36.3 K.

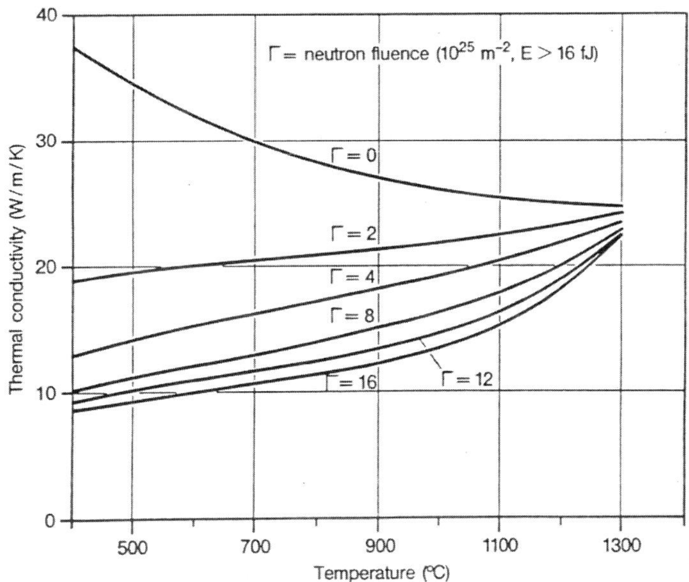

Figure 3.13. *Thermal conductivity λ of A3-3 reactor grade graphite as a function of temperature and neutron fluence. From this figure, an average value of 20 W/mK is justifiable. Reproduced from Gontard and Nabielek [1990].*

The heat conductivity of graphite depends on temperature and neutron dose (fluence), as illustrated in figure 3.13 (reproduced from Gontard and Nabielek [1990]). From the figure, $\lambda \approx 20$ W/mK seems a reasonable value for HTR graphite. In the case of the PBMR, the centerline temperature of an *average* pebble is 130 K higher than the coolant temperature, but some pebbles may see a much larger than average effective power, and thus have a steeper temperature profile. For instance in AVR pebbles, centerline temperatures up to 1400°C have been reported for an average coolant temperature of 950°C [Pohl, 2006]. For GCFR applications, the core power density will be between 50 MW/m^3 and 100 MW/m^3, i.e. 10 to 20 times higher than in this example. The modern PB-GCFR reported in Taiwo et al. [2006] uses SiC or ZrC as the matrix material. The unirradiated thermal conductivity of SiC given in Every [2006] is between 130 W/mK at 200°C and 40 W/mK at 1200°C. In that same report, the irradiated thermal conductivity of SiC is cited as 7 W/mK at 700°C and 16 W/mK at 900°. Although these numbers have large statistical errors associated with them, it can be assumed that the thermal conductivity of irradiated SiC is comparable to that of irradiated graphite and possibly worse. Using q_p = 10 kW, λ = 16 W/mK and the same pebble geometry as before results in ΔT_I = 995 K. From these considerations, it can be concluded that a pebble fuel element is unrealistic for a Gas Cooled Fast Reactor.

4

Breeding Gain for the closed nuclear fuel cycle: theory

The breeding performance of a nuclear reactor is usually formalized as the Breeding Ratio (*BR*) or Breeding Gain (*BG*). For the Generation IV GCFR breeding is not the objective, but to obtain a closed fuel cycle enough fissile material should be bred to allow refueling of the same reactor, adding fertile material to make the new fuel. In that respect, a definition of breeding gain is required taking into account all steps in the closed fuel cycle, i.e. irradiation, cool down, reprocessing, and fabrication of new fuel. In this chapter, the necessary theory is derived to calculate *BG* for the closed fuel cycle. To calculate the effect of changes in the initial fuel composition, first order perturbation theory is developed for the transmutation equation. In chapter 5, the theory is applied to the fuel cycle of a Generation IV Gas Cooled Fast Reactor.

4.1 Breeding Ratio and Breeding Gain

Intuitively the breeding performance of a reactor is found by comparing the amount of fissile material in the core at Beginning of Cycle (BOC) and End of Cycle (EOC). Now introduce the following definitions for a more detailed analysis (see Waltar and Reynolds [1981]):

FP = Fissile material produced per cycle
FD = Fissile material destroyed per cycle
FG = Fissile material gained per cycle

The *Breeding Ratio BR* is defined as the ratio of fissile production and fissile destruction:

$$BR = \frac{FP}{FD} \tag{4.1}$$

If $BR = 1$, production and destruction are exactly equal. A related parameter is the *Breeding Gain*, which is defined as the ratio of fissile material gained to fissile material destructed:

$$BG = \frac{FG}{FD} \tag{4.2}$$

The net gain of fissile material equals fissile produced minus fissile destructed, so $FG = FP - FD$, and substituting into (4.2) results in: $BG = FG/FD = (FP - FD)/FD = (FP/FD) - 1 = BR - 1$. These definitions are *integral*, i.e. they require knowledge of the fuel composition at BOC and EOC. The evolution in time of a mixture of nuclides interacting with a neutron field is determined by the so-called transmutation equation, which, for any nuclide i, is given by:

$$\frac{dN_i}{dt} = -N_i(\sigma_{a,i}\phi + \lambda_{x,i}) + \sum_j y_{j\to i}N_j\sigma_{f,j}\phi + \sum_k N_k\sigma_{c,k\to i}\phi + \sum_l \lambda_{x,l\to i}N_l + Q_i \tag{4.3}$$

In this equation, the $-N_i(\sigma_{a,i}\phi + \lambda_i)$-term describes destruction of nuclide i, either by absorption of a neutron ($\sigma_a\phi$), or by radioactive decay of mode x ($\lambda_{x,i}$). The other terms describe the production of nuclide i: as a fission product $\sum_j y_{j\to i}N_j\sigma_{f,j}\phi$, where the sum runs over all nuclides j in the system with a fission cross section, and $y_{j\to i}$ is the yield of nuclide i from fission of nuclide j. The $\sum_k N_k\sigma_{c,k\to i}\phi$-term describes production by capture, where we have to sum over all nuclides k that may produce i by capture. The last term $\sum_l \lambda_{x,l\to i}N_l$ describes production by radioactive decay from all mother isotopes l decaying to daughter product i. Finally, there may be an independent source for nuclide i, Q_i. This is a coupled system, involving all isotopes in the system, and is conveniently written as a matrix equation:

$$\frac{d\vec{N}}{dt} = \underline{M}\vec{N} + \vec{Q} \tag{4.4}$$

The transmutation matrix $\underline{M}$ contains the nuclear cross sections σ_x, decay constants λ_x, and the flux ϕ.

The multiplication factor and flux in the core are determined by the material composition of the core, and are found by solving the time-independent Boltzmann equation for a nuclear reactor [Duderstadt and Hamilton, 1976], which is repeated here for convenience:

$$\begin{aligned}
&\hat{\Omega}\cdot\nabla\phi(\vec{r},E,\hat{\Omega}) + \Sigma_t(\vec{r},E)\phi(\vec{r},E,\hat{\Omega}) = \ldots\\
&\quad\ldots + \int_0^{4\pi} d\hat{\Omega}' \int_0^{\infty} dE'\Sigma_s(\vec{r},E'\to E,\hat{\Omega}'\to\hat{\Omega})\phi(\vec{r},E',\hat{\Omega}') + \ldots\\
&\qquad\ldots + \lambda\int_0^{4\pi}\frac{d\hat{\Omega}'}{4\pi}\chi(\vec{r},E)\int_0^{\infty} dE'\nu(\vec{r},E')\Sigma_f(\vec{r},E')\phi(\vec{r},E,\hat{\Omega}')
\end{aligned} \tag{4.5}$$

Solving this equation yields the reactor eigenvalue λ and the flux as a function of space, energy and angle: $\phi(\vec{r}, E, \hat{\Omega})$. The material composition of the reactor enters into equation (4.5) through the the macroscopic cross sections $\Sigma_t(\vec{r}, E)$, $\Sigma_f(\vec{r}, E)$, and $\Sigma_s(\vec{r}, E' \to E, \hat{\Omega}' \to \hat{\Omega})$. Once the solution of the flux $\phi(\vec{r}, E, \hat{\Omega})$ is found, it can be used to make the equivalent one-group cross sections $\sigma_x(\vec{r})$, and the one-group flux $\phi(\vec{r})$ at a point $\vec{r}$ in the reactor that enter into the transmutation equation (4.3) (omitting the vector sign on r for ease of notation):

$$\bar{\sigma}_x(r) = \frac{\int_0^\infty \int_{4\pi} \sigma_x(r, E, \hat{\Omega})\phi(r, E, \hat{\Omega})d\hat{\Omega}dE}{\int_0^\infty \int_{4\pi} \phi(r, E, \hat{\Omega})d\hat{\Omega}dE} \tag{4.6}$$

and

$$\phi(r) = \int_0^\infty \int_{4\pi} \phi(r, E, \hat{\Omega})d\hat{\Omega}dE \tag{4.7}$$

The transmutation equation and Boltzmann equation are coupled: the material composition determines the flux shape $\phi(r)$, which in turn determines the evolution of the material composition through the equivalent one-group cross sections $\sigma_x(r)$ and one-group $\phi(r)$ in (4.3). When the material composition changes, the flux distribution also changes, and as a result the nuclear data for (4.3) changes. In practice the Boltzmann equation and transmutation equation are decoupled by applying a quasi-static approximation: the Boltzmann equation is solved for a given reactor composition, and using the resulting spatially dependent flux and cross sections the transmutation equation is solved over a time interval, treating the flux and equivalent one-group cross sections as constant. Then, using the newly calculated reactor composition, the Boltzmann equation is solved again, yielding an updated flux and cross sections. Thus, one allows the densities $\vec{N}(t)$ to change quicker with respect to time than the nuclear data in $\underline{M}$. This approximation was also used in the calculations of chapter 3.

Over an irradiation interval, the fuel mixture will change, and as a measure of BG (or BR) a comparison can be made of the 'reactivity' of the initial mixture and the final mixture. For this purpose, a weight w_i has to be assigned to each nuclide i, giving a measure of the contribution to the overall reactor reactivity by that nuclide. The formal parameter R is defined as:

$$R(t_0) = \langle \vec{w}(r, t), \vec{N}(r, t) \rangle \tag{4.8}$$

where brackets indicate scalar products (i.e. integration over space and time, summation over all nuclides). In this chapter it is implicitly assumed that $\vec{w}$ contains a Dirac delta function $\delta(t - t_0)$ to obtain R at t_0. If we have two values for R calculated at two points in time, using two corresponding sets of weights $\vec{w}$, the breeding gain BG can be defined as:

$$BG = \frac{R(t_2) - R(t_1)}{R(t_1)} = \frac{\langle \vec{w}_2, \vec{N}_2(r) \rangle - \langle \vec{w}_1, \vec{N}_1(r) \rangle}{\langle \vec{w}_1, \vec{N}_1(r) \rangle} \tag{4.9}$$

This definition of BG can be used between any two points in time t_1 and t_2, if correct weights $\vec{w}$ are provided. In the quasi-static approximation, we can assume that $\vec{w}(r, t)$ in (4.8) is independent of time over an irradiation interval (this approximation will be justified later).

Then taking the time derivative of (4.8) results in:

$$\frac{dR}{dt} = \langle \vec{w}(r), \frac{d\vec{N}}{dt} \rangle \tag{4.10}$$

Now introduce equation (4.3) into (4.10):

$$\frac{dR}{dt} = \sum_{i=1}^{I} w_i \{ -N_i(\sigma_{a,i}\phi + \lambda_i) + \sum_j y_{j\to i} N_j \sigma_{f,j}\phi + \sum_k N_k \sigma_{c,k\to i}\phi + \sum_l \lambda_{l\to i} N_l \} \tag{4.11}$$

where the index i runs over all isotopes and the independent source term is neglected . This expression should be integrated over the phase space of the problem. A common definition of BG is based on a simplification of equation (4.11), for instance in Salvatores [1986]:

$$BG = \frac{\sum_{i=1}^{I} w_i (N_{i-1}\sigma_{c,i-1} - N_i \sigma_{a,i})\phi}{\sum_{i=1}^{I} \Sigma_{f,i}\phi}, \tag{4.12}$$

where we have chosen to not represent the spatial integration given in Salvatores [1986]. A variation of this equation also taking into account radioactive decay, is available in the fast reactor code ERANOS [Rimpault et al., 2002]. Definition (4.12) is an example of a *differential* definition of Breeding Gain, i.e. BG is defined in terms of *reaction rates* rather than inventories. Equation (4.12) gives an instantaneous value of BG. Only the flux and cross sections at time t_0 are required to calculate BG, i.e. there is no need to calculate the evolution of the reactor materials. Note that (4.9) and (4.12) do not calculate the same property. This is illustrated in figure 4.1, where a possible time dependence of $R(t) = \langle \vec{w}, \vec{N}(t) \rangle$ is illustrated. To calculate BG according to equation (4.9) R has to be evaluated at t_1 and t_2. Definition (4.12) calculates BG from the time derivative of R, illustrated by the tangent at t_0.

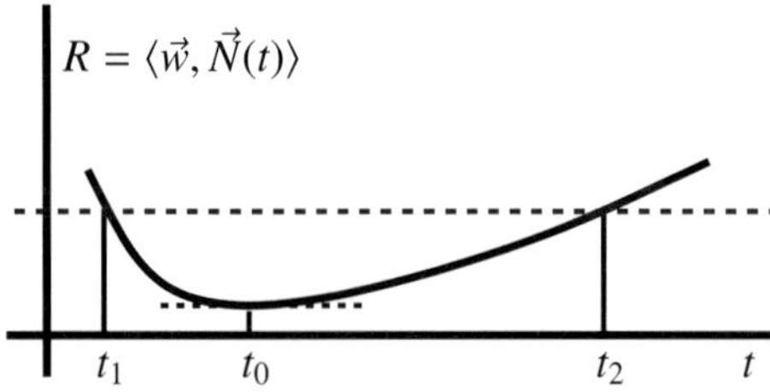

Figure 4.1. *Illustrative example of the time dependence of* $R(t) = \langle \vec{w}, \vec{N}(t) \rangle$*. Application of* (4.9) *gives* $BG = 0$ *between* t_1 *and* t_2*, whereas* (4.11) *would only give* $BG = 0$ *at time* t_0*.*

4.2 Definition of $\vec{w}$

To define $\vec{w}$, consider the change of the eigenvalue of a reactor caused by a small change of the number density of nuclide i, ΔN_i. The reference reactor is described by

$$[L_0 - \lambda_0 P_0]\phi_0 = 0, \tag{4.13}$$

with L_0, P_0 the loss and production operators ([Ott and Neuhold, 1985]). Throughout this chapter, reference values are denoted by the subscript '0', perturbed values have no subscript: $L = L_0 + \Delta L$, etc. Thus the perturbed reactor is described by:

$$[L - \lambda P]\phi = 0 \tag{4.14}$$

Using first-order perturbation theory, the eigenvalue change $\Delta\lambda$ is given by Ott and Neuhold [1985]:

$$\Delta\lambda = \frac{\langle \phi_0^*, [\Delta L - \lambda_0 \Delta P]\phi_0 \rangle}{\langle \phi_0^*, P_0 \phi_0 \rangle} \tag{4.15}$$

where ϕ_0^* is the solution of the adjoint unperturbed system. The small perturbations ΔL and ΔP can be expanded in a Taylor series, retaining only the first term:

$$\Delta L = \frac{\partial L}{\partial \alpha_i} \Delta\alpha_i \tag{4.16}$$

$$\Delta P = \frac{\partial P}{\partial \alpha_i} \Delta\alpha_i \tag{4.17}$$

in which α_i is any data element appearing in the operators L_0 and P_0. In our case, we are interested in finding the response to a change in the density if nuclide i. Substituting (4.16) and (4.17) into (4.15) for a small perturbation ΔN_i results in:

$$\Delta\lambda = \frac{\langle \phi_0^*, [\frac{\partial L}{\partial N_i} \Delta N_i - \lambda_0 \frac{\partial P}{\partial N_i} \Delta N_i]\phi_0 \rangle}{\langle \phi_0^*, P_0 \phi_0 \rangle} \tag{4.18}$$

Dividing both sides by ΔN_i results in:

$$\frac{\Delta\lambda}{\Delta N_i} = \frac{\langle \phi_0^*, [\frac{\partial L}{\partial N_i} - \lambda_0 \frac{\partial P}{\partial N_i}]\phi_0 \rangle}{\langle \phi_0^*, P_0 \phi_0 \rangle} \tag{4.19}$$

Now introduce the reactivity ρ_0, defined as $\rho_0 = 1 - \lambda_0$. Then $\Delta\rho = -\Delta\lambda$, and the reactivity weights $\vec{w}$ are defined by:

$$w_i \equiv \frac{\Delta\rho}{\Delta N_i} = \frac{\langle \phi_0^*, [\lambda_0 \frac{\partial P}{\partial N_i} - \frac{\partial L}{\partial N_i}]\phi_0 \rangle}{\langle \phi_0^*, P_0 \phi_0 \rangle} \tag{4.20}$$

Equation (4.20) is comparable to the expressions used in sensitivity and uncertainty theory, and gives a measure of the change of the reactivity of the reactor caused by density changes of an individual nuclide i. As an example, consider a reactor described in a 1-group, infinite homogeneous diffusion formalism. The operators L_0 and P_0 are given by:

$$\begin{aligned} P_0 &= \nu\Sigma_f = \sum_{i=1}^{I} N_i \nu_i \sigma_{f,i} \\ L_0 &= \Sigma_a = \sum_{i=1}^{I} N_i \sigma_{a,i} \end{aligned} \tag{4.21}$$

with the index i running over all isotopes in the system. In the 1-group formalism, ϕ_0 and ϕ_0^* reduce to single numbers. Taking the derivatives to N_i in (4.21) and substituting in (4.20) results in

$$w_i = \frac{\phi_0^*(\lambda_0 \nu_i \sigma_{f,i} - \sigma_{a,i})\phi_0}{\phi_0^* \nu\Sigma_f \phi_0} = \frac{1}{\nu\Sigma_f}(\lambda_0 \nu_i \sigma_{f,i} - \sigma_{a,i}) \tag{4.22}$$

which is similar to traditional definitions of reactivity weights as for instance found in Salvatores [1986]:

$$w_i = \nu_i \sigma_{f,i} - \sigma_{a,i} \tag{4.23}$$

The differences between (4.22) and (4.23) are the presence of the factors $1/\nu\Sigma_f$ and λ_0. These factors are not problematic, because $\lambda_0 = 1$ in a critical reactor, which is usually assumed in the derivation of (4.23), and $1/\nu\Sigma_f$ can be removed by normalizing. Note here that $\vec{w}$ depends on $\vec{N}$, and is thus implicitly time-dependent.

4.3 Reprocessing formalism

In a fuel cycle with reprocessing, a new fuel can be made using the reprocessed material, to which feed material can be added. In a closed fuel cycle, material has to be added to the fuel to offset the losses of burnup and reprocessing. In figure 4.2 a schematic is given of the fuel cycle with reprocessing. The fuel from the fuel fabrication is formally described as:

$$\vec{N}_{\text{new}} = \vec{N}_{\text{repro}} + \vec{N}_{\text{feed}} \tag{4.24}$$

with $\vec{N}_{\text{new}}$ the composition of the new fuel, $\vec{N}_{\text{repro}}$ the reprocessed material and $\vec{N}_{\text{feed}}$ the feed material. The vector $\vec{N}_{\text{repro}}$ is given by

$$\vec{N}_{\text{repro}} = \underline{S}\vec{N}_{\text{cool}} \tag{4.25}$$

with $\underline{S}$ a diagonal matrix whose elements quantify the recovery efficiency of individual isotopes ($0 \leq S_{ii} < 1$, $S_{ij,i\neq j} = 0$). $\vec{N}_{\text{cool}}$ is the material from the previous cycle after irradiation and cool down. In table 4.1, the elements S_{ii} of $\underline{S}$ are given corresponding to PUREX reprocessing with 99% efficiency, and for integral recycling with 95% efficiency. The vector of

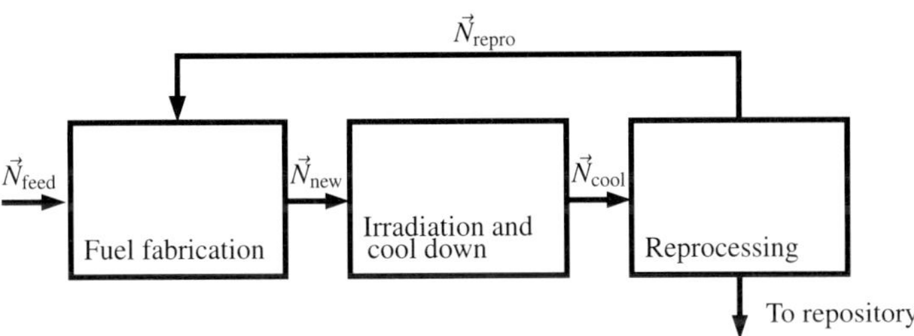

Figure 4.2. *Material flow in a reprocessing fuel cycle. The symbols are as in the text, i.e. $\vec{N}_{feed}$ is the feed material (e.g. depleted uranium), which is used together with $\vec{N}_{repro}$ to make the new fuel $\vec{N}_{new}$. After irradiation and cool down the material is given by $\vec{N}_{cool}$.*

feed material $\vec{N}_{\rm feed}$ is given by

$$\vec{N}_{\rm feed} = N_{\rm feed}.\vec{v}_{\rm feed} = [N_{\rm target} - N_{\rm repro}]\vec{v}_{\rm feed} = [N_{\rm target} - |\underline{S}\,\vec{N}_{\rm cool}|]\vec{v}_{\rm feed} \qquad (4.26)$$

Here $N_{\rm feed} = |\vec{N}_{\rm feed}|$ is the total amount of feed material. The isotopic composition of the feed material is given by the unit vector $\vec{v}_{\rm feed}$ (80% ^{238}U and 20% ^{240}Pu, for instance). $N_{\rm target}$ is the target total amount of new fuel. For multi-recycling in the same reactor, $N_{\rm target}$ equals $|\vec{N}_{\rm new}|$, i.e. the total amount of fuel at beginning of irradiation in the reactor. The amount of feed material to be added thus equals the target amount of new fuel, minus the amount of material recovered from the previous irradiation $N_{\rm repro}$. The amount of reprocessed material is given by $|\vec{N}_{\rm repro}| = |\underline{S}\,\vec{N}_{\rm cool}|$.

It is important to realize what the free parameters are in this formulation. The composition of the fuel after irradiation and cool down is determined by the burnup, neutron energy spectrum in the reactor and the length of the cool down interval. The composition of the reprocessed material is determined by the reprocessing strategy and efficiency, which are (more or less) free parameters. In a closed fuel cycle, the amount of material added is determined by the burnup of the previous irradiation and the reprocessing losses. In a truly closed fuel cycle the isotopic composition $\vec{v}_{\rm feed}$ cannot be chosen freely. The choices for $\vec{v}_{\rm feed}$ are limited to natural uranium or depleted uranium in the closed fuel cycle. Thus, the options to influence the closed fuel cycle are the burnup, cool down length, and reprocessing strategy and efficiency. Given equation (4.24), the R of the new fuel mixture can be written as:

$$R = \langle \vec{w}, \vec{N}_{\rm new}\rangle = \langle \vec{w}, \vec{N}_{\rm repro}\rangle + \langle \vec{w}, \vec{N}_{\rm feed}\rangle = R_{\rm repro} + R_{\rm feed} \qquad (4.27)$$

i.e. R of the new fuel is the sum of the R of the reprocessed material and the R of the feed material. Expanding (4.27) using (4.25) and (4.26) results in:

$$R = \langle \vec{w}, \underline{S}\,\vec{N}_{\rm cool}\rangle + [N_{\rm target} - |\underline{S}\,\vec{N}_{\rm cool}|]\langle \vec{w}, \vec{v}_{\rm feed}\rangle \qquad (4.28)$$

The R calculated for the new material using (4.28) can be compared to the R of the initial fuel, and BG can be calculated by

Table 4.1. *Examples of the reprocessing matrix $\underline{S}$, for a PUREX type reprocessing with 99% recovery efficiency, and an integral reprocessing with 95% recovery efficiency.*

Nuclide	$S_{ii,\text{PUREX}}$	$S_{ii,\text{Integral}}$
^{235}U	0.99	0.95
^{238}U	0.99	0.95
^{237}Np	0.0	0.95
^{238}Pu	0.99	0.95
^{239}Pu	0.99	0.95
^{240}Pu	0.99	0.95
^{241}Pu	0.99	0.95
^{242}Pu	0.99	0.95
^{241}Am	0.0	0.95
^{242}Am	0.0	0.95
^{242m}Am	0.0	0.95
^{244}Cm	0.0	0.95

$$BG = \frac{R_2 - R_1}{R_1} = \frac{R_2}{R_1} - 1 = \frac{\langle \vec{w}_2, \vec{N}_2 \rangle}{\langle \vec{w}_1, \vec{N}_1 \rangle} - 1 \tag{4.29}$$

where the index '1' indicates the fresh fuel at the start of the previous irradiation and '2' indicates the new fuel for the current irradiation. The set of weights $\vec{w}_2$ for the new fuel can be calculated once $\vec{N}_{\text{new}}$ is known.

4.4 Nuclide perturbation theory

It is desired to estimate the effect of variations of the initial fuel composition on the total *BG* of the fuel cycle, for instance to optimize fuel design. From the definitions of *R* and *BG* in (4.28) and (4.29) it is clear that variations of the initial fuel composition have an effect on *BG* through $\vec{N}_{\text{cool}}$. In fact, there are two effects:

1. One effect is the variation in the total amount of reprocessed material. From equation (4.28) this effect is quantified by the $|\underline{S}\,\vec{N}_{\text{cool}}|$ term.

2. The second effect is the variation of the reactivity of the reprocessed material. This term is quantified by the $\langle \vec{w}, \underline{S}\,\vec{N}_{\text{cool}} \rangle$ term in equation (4.28).

To calculate the effect of variations of the initial fuel composition on the composition of the fuel after irradiation and cool down, it is proposed to apply first order nuclide perturbation theory. In this section, the calculation of perturbations of the nuclide composition is introduced following the formulation of Williams [1986]. The transmutation equation is given by:

$$\frac{\partial \vec{N}_0}{\partial t} = \underline{M}_0 \vec{N}_0 + \vec{Q}_0 \tag{4.30}$$

with the initial condition

$$\vec{N}_0(t = 0) = \vec{N}_i. \tag{4.31}$$

Define a general response function R as:

$$R_0 = \langle \vec{b}, \vec{N}_0 \rangle \tag{4.32}$$

With $\vec{b}$ a response selection vector (for instance, $\vec{b}$ can contain $\sigma_a \phi$ at position i to give the time integrated absorption rate of nuclide i, or contain a Dirac delta $\delta(t - t_0)$ at position i to give the density of nuclide i at time t_0). This system is an example of a much broader class of functionals and responses for which a perturbation development is possible (see for instance Cacuci [2003]), if R is functional of the data and the variable governed by the functional (in this case $\vec{N}$), but $\vec{b}$ may not be dependent on $\vec{N}$. In our definition of R in equation (4.8), $\vec{w}$ takes the place of $\vec{b}$. As seen from equation (4.20), $\vec{w}$ depends on $\vec{N}$ through ϕ, ϕ^* and the normalization factor in the denominator. Neglecting the dependence of $\vec{w}$ on $\vec{N}$ introduces a first order error, but one that may be small for most variations under consideration. Assume a reference calculation $R_0 = \langle \vec{w}_0, \vec{N}_0 \rangle$ and a small perturbation giving R:

$$R = \langle \vec{w}_0 + \Delta\vec{w}, \vec{N}_0 + \Delta\vec{N} \rangle = \langle \vec{w}_0, \vec{N}_0 \rangle + \langle \vec{w}_0, \Delta\vec{N} \rangle + \langle \Delta\vec{w}, \vec{N}_0 \rangle + \langle \Delta\vec{w}, \Delta\vec{N} \rangle \tag{4.33}$$

In the first-order approximation the term $\langle \Delta\vec{w}, \Delta\vec{N} \rangle$ is neglected. Neglecting the term $\langle \Delta\vec{w}, \vec{N}_0 \rangle$ compared to $\langle \vec{w}_0, \Delta\vec{N} \rangle$ is in fact comparable to the quasi-static approximation used in burnup calculations, where it is assumed that the nuclide density $\vec{N}$ changes (much) quicker as a function of the neutron fluence ϕt than the equivalent one-group cross sections σ_x appearing in $\underline{M}_0$. The validity of the quasi-static approximation is checked in section 5.6 . To fully treat the variation in R due to $\Delta\vec{N}$, Generalized Perturbation Theory must be used. For the current study, a simplified approach is deemed instructive and sufficient.

The initial condition for (4.30) can be removed by adding the initial condition (4.31) to (4.30) as a delta function:

$$\frac{\partial \vec{N}_0}{\partial t} = \underline{M}_0 \vec{N}_0 + [\vec{N}_i \delta(t) + \vec{Q}_0] \tag{4.34}$$

with initial condition

$$\vec{N}_0(t=0) = 0 \tag{4.35}$$

The initial condition is treated as an extra term to source $\vec{Q}_0$. Now assume that we have some perturbation in the independent source: $\vec{Q} = \vec{Q}_0 + \Delta\vec{Q}$, resulting in perturbed densities $\vec{N}$. The perturbed system is given by:

$$\frac{\partial \vec{N}}{\partial t} = \underline{M}_0 \vec{N} + (\vec{Q}_0 + \Delta\vec{Q}) \tag{4.36}$$

and the corresponding perturbed response

$$R = \langle \vec{b}, \vec{N} \rangle \tag{4.37}$$

Using the general properties of adjoint equations the adjoint to (4.30) is defined:

$$\frac{-\partial \vec{N}_0^*}{\partial t} = \underline{M}_0^* \vec{N}_0^* + \vec{Q}_0^* \tag{4.38}$$

It can be shown, e.g. as in Cacuci [2003], that in order to obtain a meaningful interpretation of the adjoint equations, the adjoint boundary conditions and source must be chosen as:

$$\vec{N}_0^*(t = t_f) = 0 \tag{4.39}$$

$$\vec{Q}_0^* = \frac{\partial R}{\partial \vec{N}} = \vec{b} \tag{4.40}$$

where t_f is the final time, i.e. the forward and adjoint problems are calculated from $t_0 = 0$ to a certain final time t_f. Now form the scalar products of (4.36) with $\vec{N}_0^*$, and of (4.38) with $\vec{N}$, insert (4.40) and subtract:

$$\begin{aligned} \langle \vec{N}_0^*, \frac{\partial \vec{N}}{\partial t} \rangle &= \langle \vec{N}_0^*, \underline{M}_0 \vec{N} \rangle + \langle \vec{N}_0^*, (\vec{Q}_0 + \Delta\vec{Q}) \rangle \\ \langle \vec{N}, \frac{-\partial \vec{N}_0^*}{\partial t} \rangle &= \langle \vec{N}, \underline{M}_0^* \vec{N}_0^* \rangle + \langle \vec{N}, \vec{Q}_0^* \rangle \\ \langle \vec{N}_0^*, \frac{\partial \vec{N}}{\partial t} \rangle + \langle \vec{N}, \frac{\partial \vec{N}_0^*}{\partial t} \rangle &= \langle \vec{N}_0^*, (\vec{Q}_0 + \Delta\vec{Q}) \rangle - \langle \vec{N}, \vec{b} \rangle \end{aligned} \tag{4.41}$$

Here the adjoint property $\langle \vec{N}_0^*, \underline{M}_0 \vec{N} \rangle = \langle \vec{N}, \underline{M}_0^* \vec{N}_0^* \rangle$ is used. The LHS can be written as $\frac{d}{dt}(\vec{N}(t)\vec{N}_0^*(t))$ and integrated as indicated by the $\langle , \rangle$-operator:

$$\int_0^{t_f} \frac{d}{dt}\left(\vec{N}(t)\vec{N}_0^*(t)\right) dt = \vec{N}(t_f)\vec{N}_0^*(t_f) - \vec{N}(0)\vec{N}_0^*(0) = 0 \tag{4.42}$$

because of the boundary conditions $\vec{N}(t=0) = 0$ and $\vec{N}_0^*(t = t_f) = 0$. What remains:

$$\langle \vec{N}_0^*, (\vec{Q}_0 + \Delta\vec{Q}) = \langle \vec{N}, \vec{b} \rangle = \langle \vec{N}_0, \vec{b} \rangle + \langle \Delta\vec{N}, \vec{b} \rangle = R_0 + \Delta R$$

and thus:

$$\boxed{\Delta R = \langle \vec{N}_0^*, \Delta\vec{Q} \rangle} \tag{4.43}$$

This is an exact relation between a perturbation of the source $\vec{Q}$ and the resulting perturbation of the response R. Until now, $\vec{b}$ and $\Delta\vec{Q}$ are completely arbitrary. Remember that according to equation (4.34) a perturbation in the initial condition can be written as a perturbation to the source term:

$$\Delta\vec{Q} = \Delta\vec{N}_0\delta(t) \tag{4.44}$$

then, from (4.43), the resulting change in response at the final time t_f is given by:

$$\Delta R = [\vec{N}_0^*(t = 0)\Delta\vec{N}_0]_{\neq t} \tag{4.45}$$

where the $[]_{\neq t}$ indicates that the integration is over the entire phase space except t. From (4.45) the change in a final time response caused by a perturbation of the initial conditions can be readily calculated.

Equation (4.43) gives ΔR for a perturbation of the source $\vec{Q}$ and/or the initial condition. Another possible perturbation is a perturbation of the nuclear data appearing in the transmutation matrix $\underline{M}_0$. $\underline{M}$ will lead to a perturbed solution $\vec{N}$, but the initial conditions $\vec{N}_0(t = 0)$ are not affected. Inserting $\underline{M} = \underline{M}_0 + \Delta\underline{M}$, $\vec{N} = \vec{N}_0 + \Delta\vec{N}$ into (4.30) gives:

$$\frac{\partial}{\partial t}(\vec{N}_0 + \Delta\vec{N}) = (\underline{M}_0 + \Delta\underline{M})(\vec{N}_0 + \Delta\vec{N}) + \vec{Q}_0 \tag{4.46}$$

which is expanded to

$$\frac{\partial\vec{N}_0}{dt} + \frac{\partial\Delta\vec{N}}{dt} = \underline{M}_0\vec{N}_0 + \vec{Q}_0 + \Delta\underline{M}\vec{N}_0 + \underline{M}_0\Delta\vec{N} + \Delta\underline{M}\Delta\vec{N} \tag{4.47}$$

from which (4.30) can be subtracted to give

$$\frac{\partial\Delta\vec{N}}{\partial t} = \Delta\underline{M}\vec{N}_0 + \underline{M}_0\Delta\vec{N} + \Delta\underline{M}\Delta\vec{N} \tag{4.48}$$

Table 4.2. *1-group cross sections appearing in $\underline{M}$ for a representative GCFR fuel mixture in a closed fuel cycle. Cross sections are given for fresh and irradiated fuel (6.5% FIMA). The cross sections are almost constant during irradiation.*

	σ_a Unirrad.	σ_c Unirrad.	σ_a Irrad.	σ_c Irrad.	$\Delta\sigma_a$	$\Delta\sigma_c$
^{238}U	3.54e-1	3.00e-1	3.57e-1	3.02e-1	+0.92%	+0.63%
^{238}Pu	1.95e-0	6.19e-1	1.96e-0	6.24e-1	+0.71%	+0.77%
^{239}Pu	2.51e-0	6.09e-1	2.52e-0	6.13e-1	+0.45%	+0.78%
^{240}Pu	9.71e-1	5.64e-1	9.78e-1	5.65e-1	+0.72%	+0.3%
^{241}Am	2.33e-0	2.01e-0	2.35e-0	2.02e-0	+0.59%	+0.4%

Now form the same scalar products as in (4.41), which leads to a LHS $\frac{d}{dt}\left(\vec{N}_0^*(t)\Delta\vec{N}(t)\right)$, which can be integrated to give zero because of the boundary conditions $\vec{N}_0^*(t = t_f) = 0$ and $\Delta\vec{N}(t = 0) = 0$. Neglecting the second order term $\Delta\underline{M}\Delta\vec{N}$, ΔR becomes:

$$\boxed{\Delta R = \langle \vec{N}_0^*, \Delta\underline{M}\vec{N}_0\rangle} \tag{4.49}$$

Taking together the effects of a change in source $\vec{Q}$, initial condition $\vec{N}_0(0)$, and $\underline{M}$ gives:

$$\boxed{\Delta R = [\vec{N}_0^*(t=0)\Delta\vec{N}_0]_{\neq t} + \int_0^{t_f} \vec{N}_0^*[\Delta\vec{Q} + \Delta\underline{M}\vec{N}_0]_x dt} \tag{4.50}$$

4.5 Limitations of perturbation theory

Perturbation theory is only valid for small perturbations. The maximum allowable magnitude of a perturbation is not known a-priori, but direct non-perturbative calculations can be used to check the validity of the perturbation approach (see section 5.4). It should be noted that equation (4.43) is exact if the data appearing in $\underline{M}$ is constant. In practice however, the perturbation of the initial nuclide composition possibly gives rise to changes of the one-group cross sections appearing in the transmutation matrix $\underline{M}$. Another effect is the change of flux level and / or flux spectrum during irradiation. These effects appear as (time-dependent) changes of $\underline{M}$. Since our target is to examine the effect of perturbations of initial conditions, changes in $\underline{M}$ should be negligible. From the discussion in section 3.9, the flux level is more or less constant during irradiation in the closed fuel cycle (Σ_f is more or less constant during irradiation, hence constant power requires a constant flux level). For a representative GCFR fuel mixture, the absorption and capture cross sections were calculated for fresh fuel and at a burnup of 6.5% FIMA. The result is given in table 4.2 for the most abundant nuclides in the fuel mixture. From the numbers presented in that table, it is concluded that $\underline{M}$ does not change very much during irradiation, and that the perturbation approach is valid.

4.6 Applications of perturbation theory

In the scope of this work, we are dealing with the final time response R as defined by equation (4.27), which can be written as a sum of two final time responses involving $\vec{N}_{\text{cool}}$ (4.28). Thus, from (4.50) only the first term on the RHS remains for each response, and we find that we have to solve two adjoint calculations to correctly estimate the perturbation of R caused by perturbations to the initial nuclide inventory. The first perturbation to be considered is the 'worth perturbation' due to the reactivity of the reprocessed material R_{repro}. Since R_{repro} is given by:

$$R_{\text{repro}} = \langle \vec{w}, \underline{S} \vec{N}_{\text{cool}} \rangle \tag{4.51}$$

the corresponding 'worth adjoint' final time condition is given by:

$$\vec{N}_0^*(t = t_f) = \underline{S} \vec{w} \tag{4.52}$$

Secondly, there is an 'inventory perturbation', due to changes of the amount of feed material. This gives rise to a perturbation in R_{feed}. Since R_{feed} is given by

$$R_{\text{feed}} \propto -N_{\text{feed}} = -|\underline{S} \vec{N}| \tag{4.53}$$

the corresponding 'inventory adjoint' problem is defined by the final condition:

$$\vec{N}^*(t = t_f) = \underline{S} \vec{I} \tag{4.54}$$

with $\vec{I}$ a unity vector. Note that it is assumed that $\underline{S}$ and $\vec{v}_{\text{feed}}$, the reprocessing efficiency and vector of feed material, do not change because of the perturbation. Also note that in these calculations the flux ϕ, and not the power, is kept constant, so the final burnup also changes as a function of composition.
In principal, equation (4.50) can be used to calculate the change of BG due to changes of the nuclear data σ_x and λ_x appearing in $\underline{M}$. A small change of a cross section for instance, results in $\Delta \underline{M}$. The sensitivity of transmutation to cross section data can also be calculated using equation (4.50), by substituting $\Delta \underline{M}$ with a Taylor expansion (see also equation (4.15)):

$$S_{ij} = \frac{\Delta R_i}{R_i} \frac{\alpha_j}{\Delta \alpha_j} = \frac{\alpha_j}{R_i} \int_0^{t_f} \vec{N}_0^* \left(\frac{\partial}{\partial \alpha_j} \underline{M} \right) \vec{N}_0 dt \tag{4.55}$$

with S_{ij} the sensitivity of response i to data α_j.

4.7 Conclusions

From the developments in this chapter the following conclusions can be drawn:

- The reactivity weights $\vec{w}$ can be defined using a eigenvalue perturbation approach, equation (4.20). This definition of $\vec{w}$ reduces to a commonly used definition (equation (4.23)) if the reactor is described in a 1-group, infinite homogeneous medium formalism.

- Breeding Gain can be defined as an integral parameter, based on nuclide inventories at BOC and EOC, or as a differential parameter, based on reaction rates. If the reactivity weights $\vec{w}$ are treated as constants (quasi-static approximation), it can be shown that the differential *BG* at time t_0 (e.g. (4.12)), corresponds to the time derivative of the integral parameter R (equation (4.10) sqq.). An integral definition of *BG*, such as equation (4.9) is more accurate, but requires solving an entire irradiation history, whereas the differential *BG* can be readily calculated if nuclide composition and flux spectrum are known.

- For the closed fuel cycle, the definition of *BG* should include all steps between BOC of fuel batch i and BOC of fuel batch $i+1$, where fuel batch $i+1$ is made using reprocessed material from batch i. In this chapter, such a definition is given in equations (4.28) and (4.29). If the reactivity weights are treated as constants, first order perturbation theory can be applied to estimate the effect on *BG* of variations of the initial fuel composition (section 4.6).

5

Breeding gain for the closed nuclear fuel cycle: application

In this chapter the theory that was developed in chapter 4 is applied to the fuel cycle of GFR600, a 600 MWth Gas Cooled Fast Reactor. This reactor is envisaged to run in a closed fuel cycle [U.S. DOE Nuclear Energy Research Advisory Committee and the Generation IV International Forum, 2002], breeding just enough fissile material to refuel the reactor, only adding fertile material to the reprocessed material. All actinides are recycled indiscriminately in this concept. The GFR600 reactor is investigated as part of the European 6th Framework Program GCFR-STREP (*S*pecific *T*argeted *RE*search *P*rogram). The theory is applied to the fuel cycle of this reactor to determine *BG* for several fuel concepts and reprocessing options.

5.1 Reactor model and calculational tools

All calculations presented in this chapter are based on unit cell calculations. This calculational approach is justified as follows: almost every nuclear reactor uses some kind of fuel management, where fuel assemblies are shuffled from time to time. This gives rise to different irradiation conditions over the irradiation cycle. In the case of the closed fuel cycle, the effect of zoning will generally be small. As indicated in section 3.9, the macroscopic fission cross section Σ_f of the fuel is preserved or increases during irradiation in a closed fuel cycle. Assuming that all nuclides have a similar energy release per fission, this means that the flux level required for a predefined power production, is rather constant as a function of burnup. The buildup of fission products during irradiation causes a spectral effect during burnup. For fast reactors, this effect is not very significant. Zoning of fuel in the closed fuel cycle may be somewhat arbitrary, because reactivity of the fuel does not change very much during irradiation. It is therefore assumed that cell calculations can give adequate insight in the physics of the GFR600 fuel cycle. Some key parameters of the reactor are given in table 5.1. The fuel for this reactor is a plate fuel, using a fuel composed of a mix of (U, Pu, MA)C (70 vol%) and SiC matrix (30 vol%), clad by thin layers of SiC. The GFR600 fuel assembly and a detail of the fuel slabs are given in figure 5.1. Table 5.2 gives the composition of the reference fuel and a fuel containing 5% MA (Np, Am, Cm). Because an integral fuel cycle is envisaged for GFR600 the influence of adding MA on the breeding gain is investigated.

Table 5.1. *GFR600 core parameters.*

Power [MWth]	600
Coolant	He
Power density [MW/m^3]	103
Specific power [W/gHM]	45
$T_{core,in}$ [°C]	480
$T_{core,out}$ [°C]	850
Core H/D [m/m]	1.95/1.95
Pressure [MPa]	7.0
Fuel type	plates
Fuel material	UPuC + MA
Struct. material	SiC
Refl. material	Zr_3Si_2
Vol.% coolant / structures / fuel	55/10/35

The weights $\vec{w}$ are obtained using the sensitivity module TSUNAMI-1D of the SCALE 5 code system [Oak, 2005]. TSUNAMI-1D calculates the sensitivity of k_{eff} to the nuclear data, based on a unit cell calculation on an infinite lattice with fuel, cladding and moderator, in a slab, cylindrical or spherical configuration. Axial leakage is taken into account by a buckling correction. TSUNAMI-1D calculates the sensitivity coefficients of k_{eff} to the nuclear data in the following way ([Broadhead et al., 2004]):

$$S_{k_{\text{eff}},\Sigma_x} \equiv \frac{\partial k_{\text{eff}}}{k_{\text{eff}}} / \frac{\partial \Sigma_x}{\Sigma_x} = -\frac{\Sigma_x}{k_{\text{eff}}} \frac{\langle \phi_0^*, [\frac{\partial L}{\partial \Sigma_x} - \frac{1}{k_{\text{eff}}} \frac{\partial P}{\partial \Sigma_x}]\phi_0 \rangle}{\langle \phi_0^*, \frac{P}{k_{\text{eff}}^2} \phi_0 \rangle} \tag{5.1}$$

with Σ_x a nuclide reaction cross section. The sensitivities have an explicit component and an implicit component. The implicit component is due to the effect of perturbations of a reaction cross section on the self-shielding of other cross sections. The implicit component is important for instance for ^{238}U self-shielding in water moderated lattices. TSUNAMI-1D treats both implicit and explicit effects and calculates the total sensitivity. For nuclide i, the variation of the total cross section $d\Sigma_{t,i}$ equals:

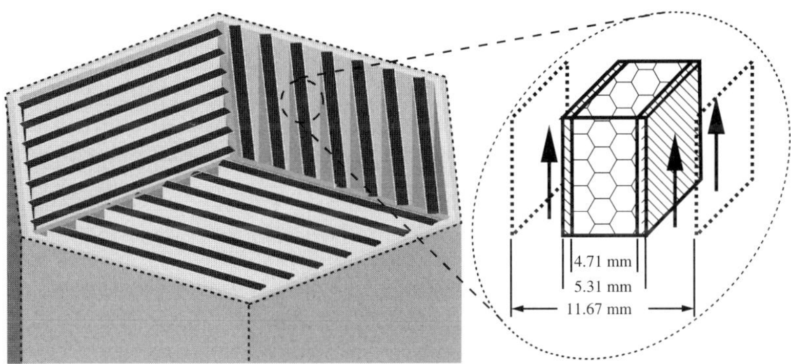

Figure 5.1. *GFR600 fuel assembly. The light colored wrapper and central mechanical restraint are made of SiC. The darker fuel slabs contain the fuel mixture clad with SiC. The fuel thickness is 4.71 mm, cladding thickness 0.3 mm, and pitch 11.67 mm. The fuel material is a mix of 30 vol% matrix (SiC), and 70 vol% of a UPuC + MA mixture with 15 % porosity. Each assembly contains 21 fuel plates. The fueled length is 1.95 m, the overall length of the assembly about 4.4 m. The overall volume fractions are 55 vol% helium, 10 vol% SiC for structures and cladding and 35 vol% fuel/matrix.*

$$d\Sigma_{t,i} = N_i d\sigma_{t,i} + \sigma_{t,i} dN_i, \tag{5.2}$$

so $d\Sigma_{t,i}$ can be interpreted as either the sensitivity to the nuclear data at constant number density N_i, or the sensitivity to the density at constant nuclear data $\sigma_{t,i}$. Substituting the RHS of (5.2) into (5.1) gives:

$$S_{k_{\text{eff}},\Sigma_t} = -\frac{N_i\sigma_{t,i}}{k_{\text{eff}}}\frac{\langle\phi_0^*, \frac{1}{\sigma_{t,i}}[\frac{\partial L}{\partial N_i} - \frac{1}{k_{\text{eff}}}\frac{\partial P}{\partial N_i}]\phi_0\rangle}{\langle\phi_0^*, \frac{P}{k_{\text{eff}}^2}\phi_0\rangle} \tag{5.3}$$

Comparing this to (4.20) shows that the weight w_i can be calculated from (5.3):

$$w_i = S_{k_{\text{eff}},\Sigma_t}\frac{k_{\text{eff}}}{N_i} \tag{5.4}$$

allowing for easy evaluation of w_i. TSUNAMI-1D returns the total $S_{k_{\text{eff}},\Sigma_t}$, i.e. the implicit component is taken into account. For GFR600, the calculations have shown that for all actinides, the implicit component is 2 to 3 orders of magnitude smaller than the explicit component. The approach of equations (5.3) and (5.4) is recommended in the TSUNAMI-1D manual as a way to compare the perturbative calculations with direct calculations, i.e. perform 2 direct calculations and calculate the sensitivity to $\sigma_{t,i}$ as $\Delta k_{\text{eff}}/\Delta N_i$. In our case, we reverse the argument, and take $S_{k_{\text{eff}},\Sigma_t}$ as a measure of Δk_{eff} brought on by a change ΔN_i.

An adjoint-capable burnup calculation code was developed based on the ORIGEN-S formalism. This code is called LOWFAT: *L*ike *O*rigen *W*ith *F*orward and *A*djoint *T*ransmutation,

Table 5.2. *GFR600 fuel compositions: reference fuel, and a fuel containing 5% MA.*

		Reference fuel		MA fuel
		Elemental composition		
U		84%		79%
Pu		16%		16%
MA		-		5%
		Isotopic composition		
	at. dens. $[b.cm]^{-1}$	at. %	at.dens. $[b.cm]^{-1}$	at. %
^{235}U	1.1631e-4	0.60	1.0938e-4	0.56
^{238}U	1.6291e-2	83.52	1.5320e-2	78.60
^{237}Np	-	-	1.6242e-4	0.83
^{238}Pu	8.3625e-5	0.43	8.3625e-5	0.43
^{239}Pu	1.7395e-3	8.92	1.7395e-3	8.92
^{240}Pu	8.0163e-4	4.11	8.0163e-4	4.11
^{241}Pu	2.2940e-4	1.18	2.2940e-4	1.18
^{242}Pu	2.2235e-4	1.14	2.2235e-4	1.14
^{241}Am	2.1687e-5	0.11	6.0680e-4	3.11
^{242m}Am	-	-	2.2072e-6	0.01
^{243}Am	-	-	1.5020e-4	0.77
^{242}Cm	-	-	1.9219e-7	0.001
^{243}Cm	-	-	7.0178e-7	0.004
^{244}Cm	-	-	4.9558e-5	0.254
^{245}Cm	-	-	1.2086e-5	0.062
^{246}Cm	-	-	9.4528e-7	0.005

and it is described in more detail in appendix A. LOWFAT calculates the depletion, cool down and reprocessing of the HM mixture. The code uses problem dependent nuclear cross sections in the ORIGEN-S format, prepared by the CSAS and COUPLE modules of the SCALE 5 system. All calculations were done over a period of 1300 days irradiation at a constant flux $\phi = 2.10^{15}$ n/cm^2.s^1, leading to a burnup of roughly 6.5 % FIMA. The constant flux approach is justified by noting that in a closed fuel cycle, constant flux is roughly yields a constant power production during irradiation (section 3.9). After irradiation a cool down period of 6 years (2192 days) was assumed. Reprocessing is assumed to take a small amount of time compared to the cool down period. Reprocessing is assumed to be PUREX (only U, Pu recycled) or integral, where U, Np, Pu, Am and Cm are recycled. An example of the corresponding elements S_{ii} of $\underline{S}$ for PUREX with 99% efficiency, and integral recycling with 95% efficiency is given in table 4.1.

5.2 Comparison with existing definitions of $\vec{w}$

To check the consistency of our definition of $\vec{w}$ (equation (4.20)) with existing definitions of $\vec{w}$ some reactivity weights were calculated for GFR600 and compared to LMFBR values from Salvatores [1986]. The result is given in table 5.3. The weights in the table are normalized as follows:

$$w_i = \frac{w^+ - w^8}{w^9 - w^8} \tag{5.5}$$

where w^+ is the weight of a nuclide calculated with equation (4.23), and w^8, w^9 the weights of ^{238}U and ^{239}Pu respectively. This definition results in a weight of 0.0 for ^{238}U and 1.0 for ^{239}Pu. The weights calculated for GFR600 using the new definition of equation (4.20) are similar to the ones cited in Salvatores [1986], giving confidence that our new definition is consistent with existing definitions.

Table 5.3. *Weights for various nuclides from Salvatores [1986], and calculated for GFR600 using equation (4.20). The results are similar, giving confidence that the new definition for $\vec{w}$ is consistent with existing definitions. The GFR600 fuel with MA has a harder spectrum, which is evidenced by the change of weights: the threshold fissioners, like ^{240}Pu, become somewhat more important. Fuel compositions according to table 5.2.*

	From Salvatores [1986]	GFR 600 (ref)	GFR600 (5% MA)
^{235}U	0.777	0.810	0.799
^{238}U	0.000	0.000	0.000
^{239}Pu	1.000	1.000	1.000
^{240}Pu	0.130	0.144	0.175
^{241}Pu	1.542	1.520	1.457
^{242}Pu	0.032	0.093	0.120

In table 5.4 we present the weights of various isotopes calculated from equation (5.4). All weights are normalized to ^{239}Pu. Note that ^{238}U, ^{237}Np, and ^{241}Am have negative weights ($\sigma_a > \nu\sigma_f$). Also note that the weight of ^{238}Pu in the GFR600 spectrum is comparable to ^{235}U. The fuel with MA has a harder spectrum, which is evidenced by the change of weights: the threshold fissioner ^{240}Pu becomes somewhat more important.

5.3 Illustrations of adjoint transmutation calculations

In this section two illustrative examples of adjoint calculations are given in figures 5.2 and 5.3. In appendix A more background is given on the physical interpretation of the nuclide adjoints. Figure 5.2 shows three adjoints for ^{241}Pu. Consider the solid line, being the *'inventory adjoint'*, i.e. the adjoint quantifying changes of the amount of reprocessed material due to ^{241}Pu in an *integral* fuel cycle. It is the solution of the adjoint equation for equations

Table 5.4. *Weights for various nuclides calculated from equation (4.20), normalized to ^{239}Pu. The GFR600 reference fuel description does not include ^{237}Np, $^{242(m)}$Am and ^{244}Cm, so no weights were calculated for these isotopes. Notice that the weight of ^{238}Pu is quite large, and that ^{241}Am is a very strong absorber (large negative weight). In the somewhat harder spectrum of the fuel with 5% MA, the threshold fissioners like ^{238}U have a higher weight.*

	GFR 600 (ref)	GFR600 (5% MA)
^{235}U	0.794	0.785
^{238}U	-0.079	-0.067
^{237}Np	-	-0.163
^{238}Pu	0.639	0.659
^{239}Pu	1.000	1.000
^{240}Pu	0.077	0.120
^{241}Pu	1.560	1.486
^{242}Pu	0.026	0.063
^{241}Am	-0.368	-0.222
^{242}Am	-	2.140
^{242m}Am	-	2.104
^{244}Cm	-	0.213

(4.53) and (4.54) with $\underline{S}$ for integral recycling. In the *integral* recycling scheme, reprocessed material due to ^{241}Pu is the plutonium itself, plus any daughter products of ^{241}Pu that are also recycled. E.g., in the integral reprocessing scheme, the daughter product ^{241}Am is also reprocessed material due to ^{241}Pu, while in PUREX reprocessing scheme it is not.

During cool down (the right hand part of the graph), the inventory adjoint for ^{241}Pu is flat: ^{241}Pu, as well the daughter product ^{241}Am are recycled. Thus, adding ^{241}Pu during the cool down period will always lead to more reprocessed material, either as ^{241}Pu or ^{241}Am. In the *PUREX* reprocessing on the other hand, only U and Pu are recycled. If ^{241}Pu were to be added at the beginning of the cool down period, some of it will decay before reprocessing. Since the daughter product is not recycled in this scheme, the inventory adjoint for ^{241}Pu becomes time dependent, as given by the dotted line, which gives the inventory adjoint for ^{241}Pu in a *PUREX* fuel cycle.

For both reprocessing schemes, the inventory adjoint is time dependent during irradiation: if ^{241}Pu is added at the beginning of the irradiation period, the probability of fission during the irradiation is quite large. Thus, ^{241}Pu added to the fresh fuel will have a small contribution to the total amount of reprocessed material, and hence it will have a low inventory adjoint at the beginning of the irradiation interval.

In figure 5.2 the *'worth adjoint'* of ^{241}Pu is given by the dashed line. It is the solution of the adjoint according to equations (4.51) and (4.52). This line represents the contribution to the reactivity of the reprocessed material by ^{241}Pu in the integral reprocessing scheme. During

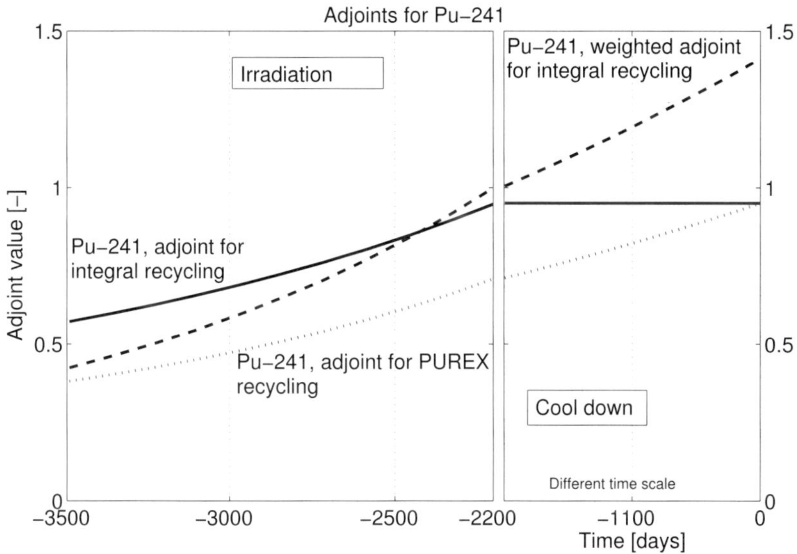

Figure 5.2. *Three adjoints for ^{241}Pu. The adjoint density is a measure of the contribution of ^{241}Pu to the amount of reprocessed material, while the weighted adjoint density gives the contribution of ^{241}Pu to the reactivity of the reprocessed material. Note the time axis: the cycle length is 1300 days of irradiation and 2192 days of cool down. Since the adjoint is solved backward in time, we are 'looking back' in time. The time scale changes at -2200 days, because both forward and adjoint densities change slowly during cool down.*

irradiation and cool down this adjoint is time dependent, because in both cases ^{241}Pu has a probability of not surviving until reprocessing, which implies that it does not contribute to the reactivity of the new fuel.

In figure 5.3 the *worth* adjoints for some absorbers are illustrated, namely ^{238}U, ^{237}Np and ^{241}Am. These lines correspond to the adjoint using equations (4.51) and (4.52). The worth adjoint represent the contribution to the reactivity of the reprocessed material due to the isotopes in question. Since these nuclides all have long half lives, the worth adjoints during cool down are constant. During irradiation the worth adjoints are time dependent. ^{237}Np and ^{241}Am have a positive worth adjoint value at the beginning of the irradiation. This may seem strange, because these nuclides have negative weights. During irradiation ^{237}Np transmutes to ^{238}Pu and ^{241}Am to $^{242(m)}$Am. These products of transmutation have positive weights. Thus, if ^{237}Np and ^{241}Am are added at the beginning of the irradiation, they have enough time to capture a neutron, after which their transmutation products contribute positively to the reactivity of the new fuel made from the reprocessed material. If ^{237}Np and ^{241}Am were to be added at a later stage of the irradiation, the probability of transmutation is smaller, and their overall contribution to the reprocessed fuel would be negative. The behavior of the worth ad-

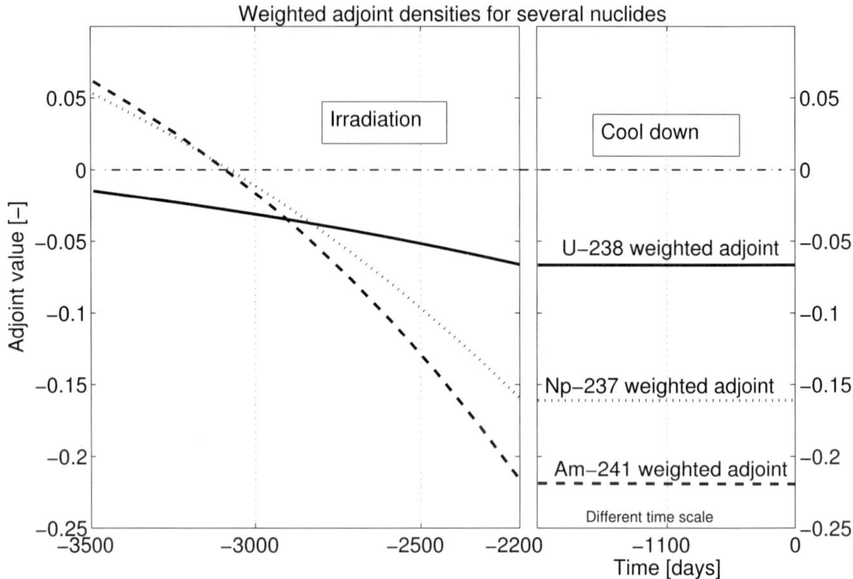

Figure 5.3. Weighted adjoints for the absorbing nuclides ^{238}U, ^{237}Np, and ^{241}Am, i.e. the contribution of these nuclides (by themselves or daughter isotopes) to the reactivity of the reprocessed fuel. Notice that ^{237}Np, and ^{241}Am have a positive contribution if they are added at the beginning of the irradiation, because they transmute to fissile isotopes. The adjoint of ^{238}U remains negative. If the irradiation time would be longer, i.e. if the final burnup is higher, more ^{238}U will have transmuted and its contribution will become positive. The time axis corresponds to a cycle of 1300 days of irradiation and 2192 days of cool down. The time scale changes from irradiation to cool down (-2200 days) because both forward and adjoint densities change slowly during cool down.

joint for ^{238}U has a similar behavior, but is never positive. This means that under the chosen irradiation conditions, the transmutation to ^{239}Pu does not have enough time to take place on a large enough scale. Here one immediately sees the influence of the final fluence (burnup) on the fuel cycle, because if the final fluence would be larger, more ^{238}U would transmute to ^{239}Pu and the reactivity worth of the reprocessed material would be larger.

5.4 Breeding Gain calculations on the GFR600 fuel cycle

One cycle of irradiation, cool down and reprocessing was calculated for the GFR600 reference fuel, and for a fuel containing 5% MA. The reprocessing scheme is integral with 99% efficiency, adding depleted uranium to make the new fuel. Refer to table 5.5: R_{ini} is R for the fresh fuel, R_{irrad} is at the end of the irradiation (i.e. after 1300 days), R_{cool} is after 6 years of cool down, and R_{repro} is the reactivity of the reprocessed material. The new fuel is composed of reprocessed and feed material (see equation (4.27)). R of the feed material is given by R_{feed} and the R of the new fuel is $R_{\text{new}} = R_{\text{repro}} + R_{\text{feed}}$. R_{feed} is dominated by the weight of ^{238}U, which is negative. With the reference fuel, BG over the irradiation ($R_{\text{irrad}}/R_{\text{ini}} - 1$) is close to zero, but during cool down reactivity is lost, resulting in an overall BG from the initial fuel to the new fuel of -20.2%. For the MA fuel BG is positive during the irradiation (+ 12%), effectively offsetting the loss during cool down, resulting in an overall BG close to zero (-3.56%). Concluding: adding a small amount of MA to the fuel results in a positive BG over the irradiation. Over the entire cycle, BG is close to zero.

Table 5.5. *Evolution of R for the reference fuel and a fuel with MA. All values are 10^{-3} unless stated otherwise. R_{ini} = R of initial fuel mixture, R_{irrad} is after irradiation, R_{cool} is after 2192 days of cool down, R_{repro} is the R of the reprocessed material. The reactivity of the feed material is given by R_{feed}, and the new fuel is described by R_{new}.*

	GFR 600 (ref)	GFR600 (5% MA)
R_{ini}	1.02	1.15
R_{irrad}	1.01	1.29
R_{cool}	0.932	1.22
R_{repro}	0.923	1.21
R_{feed}	-0.109	-0.097
R_{new}	0.813	1.11
BG	-20.2 %	-3.56 %

For the MA fuel adjoints were calculated corresponding to 99% efficient integral recycling to determine the influence of initial variations on BG. To get BG closer to zero, two perturbations were considered, namely:

1. Increase of 5% of ^{235}U, ^{239}Pu, ^{240}Pu and ^{241}Pu
2. Increase of 5% of ^{238}U

The reason for this choice is that is possible to increase the reactivity of the reprocessed fuel by starting with a larger fissile content (option 1), or by increasing the total amount of reprocessed material, thereby reducing the amount of (negative reactivity from) feed material (option 2).

The results are given in table 5.6, where the symbols are as before. The table gives the values of R for the reference calculation (same as table 5.5), after which the ΔR introduced by the perturbations are given. E.g. adding ^{235}U has a positive ΔR_{repro}: the reactivity of the reprocessed material is higher. ^{235}U also has a positive ΔR_{feed}: more ^{235}U means a smaller amount of feed material is added to make the new fuel, and since the weight of the feed material is negative, adding less of it has a positive effect. For the other isotopes similar arguments apply. The perturbation of ^{238}U (right column) has a negative effect on R_{repro} but a positive effect on R_{feed}.

Table 5.6. *Results of 2 perturbations on the evolution of R. The reference situation is a 5% MA fuel. 2 perturbations are applied and their effects calculated. 2 forward calculations were done to check the result. All values are 10^{-3} unless stated otherwise.*

		R_{ini}	1.15		
		R_{irrad}	1.29		
		R_{cool}	1.22		
		R_{repro}	1.21		
		R_{feed}	-0.097		
		R_{new}	1.11		
		BG	-3.56%		
ΔN	ΔR_{repro}	ΔR_{feed}	ΔN	ΔR_{repro}	ΔR_{feed}
^{235}U	2.39e-6	2.39e-7	^{238}U	-1.14e-5	4.92e-5
^{239}Pu	5.18e-5	3.86e-6			
^{240}Pu	6.71e-6	2.34e-6			
^{241}Pu	5.08e-6	4.55e-7			
Σ	6.60e-5	6.90e-6	Σ	-1.14e-5	4.92e-5
R'_{new}		1.19	R'_{new}		1.15
	Forward			*Forward*	
R_{ini}		1.27	R_{ini}		1.11
R_{irrad}		1.36	R_{irrad}		1.28
R_{cool}		1.29	R_{cool}		1.21
R_{repro}		1.28	R_{repro}		1.20
R_{feed}		0.09	R_{feed}		0.048
R_{new}		1.19	R_{new}		1.15
BG		- 6.42%	BG		+ 3.71 %

To check the validity of the perturbation calculations, and as a check of the employed algorithms, two non-perturbative calculations were done, i.e. forward calculations with the perturbed initial vectors. The result of these calculations is given in the two bottom cells of each column of table 5.6 The forward calculations yield the same R_{new} as the perturbation calculations. The forward calculations give all values of R during the irradiation and cool down with the perturbed nuclide inventory.

Perturbing the fissile content yields a higher R_{new} but the initial R_{ini} is also higher: the initial mixture is more reactive. The extra initial reactivity is so large, that BG is in fact worse: from -3.56 % to -6.42%. The perturbation of ^{238}U yields a lower initial R_{ini}, and a better BG: +3.71%. The perturbation of ^{238}U can be calculated explicitly: writing out the equations for a perturbation in ^{238}U:

$$\Delta R_{repro} = N^*(0)\Delta N(0) = -1.4849 \cdot 10^{-2}\Delta N(0) \tag{5.6}$$

with $N^*(0)$ the *worth* adjoint at time $t = 0$ for ^{238}U (see figure 5.3). For ΔR_{feed} we find:

$$\Delta R_{\text{feed}} = N^*(0)\langle\vec{w}, \nu_{\text{feed}}\rangle\Delta N(0)$$
$$= (0.9669)(6.6423\cdot 10^{-2})\Delta N(0) = 6.4225\cdot 10^{-2}\Delta N(0) \quad (5.7)$$

with $N^*(0)$ the *inventory* adjoint value at $t = 0$ for ^{238}U, and $\Delta N(0)$ the perturbation in the initial concentration of ^{238}U. It is seen that a perturbation in ^{238}U has a much stronger positive effect through the amount of feed material ($\Delta R_{\text{feed}} = 0.064\Delta N(0)$), than the negative effect on the reprocessed material ($\Delta R_{\text{repro}} = -0.015\Delta N(0)$): under the given irradiation circumstances, adding extra ^{238}U gives a better new fuel, but the reactivity of the initial fuel is lower.

5.5 Long term behavior of breeding gain and k_{eff}

To investigate the long-term behavior a series of 15 irradiation, cool down and reprocessing cycles was calculated. Since the cycle length is some 9.5 years, 15 cycles is longer than reactor lifetime, but it serves to illustrate the long-term behavior. Two reprocessing strategies are reported here:

1. Integral reprocessing with 99% efficiency, after which depleted uranium is added.
2. Integral reprocessing at 95% efficiency, after which a mix of 90% depleted uranium and 10% MA (same vector as before) is added.

The choice for the second strategy is given by the fact that R_{repro} is lower if S_{ii} is smaller (i.e. reactivity is lost due to large reprocessing loss), leading to the necessity to increase R_{feed} to make the overall *BG* closer to zero. In figure 5.4 the evolution of the k_{eff} of the fresh fuel and the reactivity weight R_{new} are illustrated. Both fuel cycle strategies converge to a situation where k_{eff} and R_{new} of the fresh fuel are more or less equal from cycle to cycle, so we approach the required situation of a zero *BG* over the entire fuel cycle.
For the 99% efficient reprocessing, *BG* (equation (4.29)) during irradiation is about +15%. The 95% efficient strategy, which adds some MA to the new fuel, has a *BG* between +15% and +25% during irradiation to offset the losses of reprocessing. In table 5.7 the fuel composition for the 10th cycle is given. In both fuel cycles the MA loading of the fuel remains limited. Of the two presented strategies, the first strategy (refueling with reprocessed material and DU only) would be considered as a Generation IV fuel cycle: enough new fissile material is bred during the cycle to allow refueling with a fertile material only. The other strategy requires an external source of MA, but no Pu. Note: The amount of feed material is the same as the burnup reached, so for 6.5% FIMA, the amount of feed material is 6.5% of the core inventory. Adding material with 10% MA means that the overall amount of added MA equals 0.65% of the total core inventory ($\approx$ 105 kg, the core inventory being some 16 tons of HM).
We can thus draw the conclusion that a truly closed fuel cycle is possible if the reprocessing losses are small enough (the 99% strategy). If the reprocessing scheme is less efficient in recovering the actinides, *BG* can be made closer to zero by adding depleted uranium with a

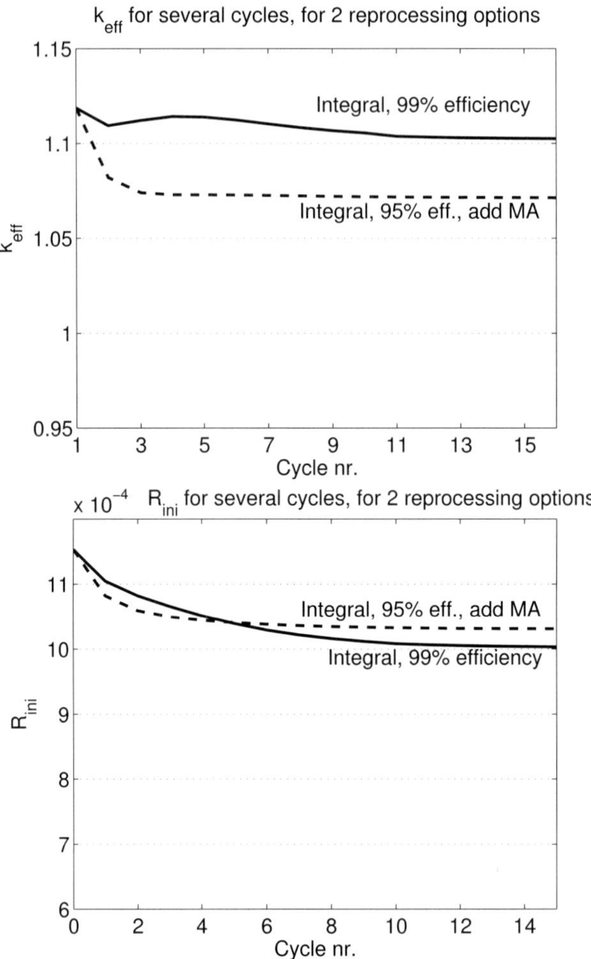

Figure 5.4. *Evolution of k_{eff} and R during multiple recycles for two fuel cycle schemes: 99% efficiency refers to an integral fuel cycle with 99% reprocessing efficiency, adding depleted uranium to make the new fuel. 95% efficiency refers to the scheme with 95% efficient reprocessing, after which a mix of depleted uranium and MA is added. Both schemes converge to a situation where the k_{eff} and the R_{new} are constant from cycle to cycle. Since R_{new} is the same from cycle to cycle, BG = 0.*

small amount (some 10 %) of MA. Transmutation of the MA will then offset the reactivity losses during reprocessing. Due to the low fissile enrichment (see table 5.7), the k_∞ of the fuel is not very high.

Table 5.7. *Fuel composition at the start of the tenth irradiation cycle for two different reprocessing strategies. Strategy 1 assumes 99% efficient reprocessing after which depleted uranium is added. Strategy 2 assumes 95% efficient reprocessing, after which depleted uranium and 10% MA are added. All numbers are in atom%.*

Isotope	Strategy 1		Strategy 2	
U		82.512%		80.78%
^{234}U	0.201%		0.331%	
^{235}U	0.072%		0.091%	
^{236}U	0.105%		0.086%	
^{238}U	81.933%		80.272%	
Np		0.119%		0.588%
^{237}Np	0.119%		0.588%	
Pu		16.513%		15.355%
^{238}Pu	0.457%		1.173%	
^{239}Pu	9.258%		8.410%	
^{240}Pu	5.607%		4.648%	
^{241}Pu	0.489%		0.384%	
^{242}Pu	0.702%		0.740%	
Am		0.838%		2.709%
^{241}Am	0.577%		1.952%	
$^{242(m)}$Am[a]	0.038%		0.106%	
^{243}Am	0.223%		0.651%	
Cm		0.219%		0.568%
^{243}Cm	0.003%		0.007%	
^{244}Cm	0.153%		0.419%	
^{245}Cm	0.033%		0.095%	
^{246}Cm	0.023%		0.038%	
^{247}Cm	0.005%		0.007%	
^{248}Cm	0.002%		0.002%	

[a] ^{242}Am and ^{242m}Am summed

5.6 Checking the validity of the quasi-static approximation for $\vec{w}$

With the results of the 15 cycles of irradiation, cool down and reprocessing, the validity of equation (4.29) and of the quasi-static approach in section 4.1 can be checked. To this end, the *BG* was calculated for both reprocessing strategies using two slightly different definitions.

The first definition is:

$$BG = \frac{\langle \vec{w}_i, \vec{N}_{i+1} \rangle}{\langle \vec{w}_i, \vec{N}_i \rangle} - 1 \tag{5.8}$$

with i the cycle number: the same set of weights $\vec{w}_i$ is used to calculate the reactivity of the fuel of batch i and the new fuel of batch $i+1$. This is the quasi-static calculation: the weights are kept constant while the nuclide vectors are allowed to vary.
The second definition calculates BG using weights $\vec{w}_i$ of the fuel in batch i, and weights $\vec{w}_{i+1}$ for the new fuel in batch $i+1$:

$$BG = \frac{\langle \vec{w}_{i+1}, \vec{N}_{i+1} \rangle}{\langle \vec{w}_i, \vec{N}_i \rangle} - 1 \tag{5.9}$$

If the weights $\vec{w}$ change considerably between cycle i and cycle $i+1$, the two definitions would result in different values of BG. If both BG definitions give comparable results, it shows that $\vec{w}$ does not change too much from cycle to cycle and that the quasi-static approach is valid. The result is given in figure 5.5. The BGs from one cycle to the next are all close to zero. It is concluded that since the BGs calculated with the two equations given above are comparable, the weights $\vec{w}$ do not change much between the cycles, and that the quasistatic approach is valid for the closed fuel cycle with integral recycling. An explanation is needed for the BG calculated for the first cycles, which are all much lower than the BGs of the later cycles. This effect is contributed to the fact that the initial fuel material has a relatively large amount of non-fertile material. Compare table 5.2, column 5, and table 5.7. The initial fuel with 5% MA contains only 78.6% ^{238}U, whereas the equilibrium value is around 81%. The initial fuel lacks fertile ^{238}U, and thus looses reactivity during the first cycle(s), resulting in a negative BG. In the later cycles, an equilibrium is obtained and BG approaches zero.

5.7 Conclusions

In this chapter, the theoretic framework developed in chapter 4 was applied to the fuel cycle of GFR600. The results lead to the following conclusions:

- The proposed definition of $\vec{w}$ based on eigenvalue perturbation theory (equation (4.20)), correctly treats self-shielding, and can be used with different formalisms (diffusion, transport, 1-group, multi-group etc). It is an improvement over existing definitions. Application of the definition yields numerical values which are in line with the existing definition of $\vec{w}$, giving confidence that the new definition is reasonable.

- Using the proposed definition of BG (equation (4.9)) for the closed fuel cycle, $BG = 0$ (from BOC of one cycle to BOC of the next) does correspond to k_{eff} being roughly the same at BOC for the fresh and the reprocessed fuel.

- The quasistatic approximation where the reactivity weights $\vec{w}$ are treated as constants, is in fact acceptable, as shown in section 5.6. This indicates that application of first

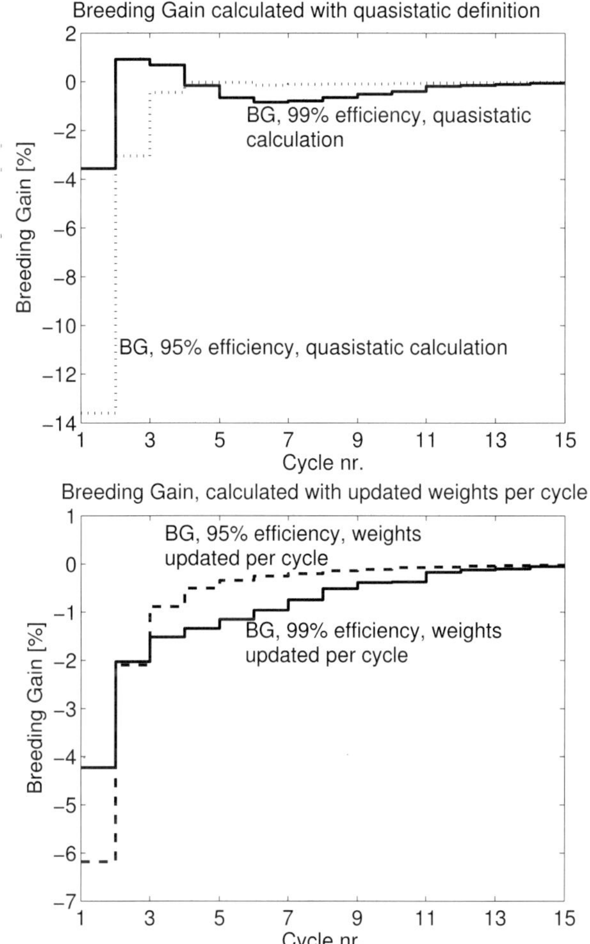

Figure 5.5. *Values of BG calculated according to equations (5.8) and (5.9) for the 99% efficient and 95% efficient reprocessing strategies. The low values in the earlier cycles are explained in the text.*

order perturbation theory (FOPT) is possible, because FOPT requires, amongst others, the weights to be constant.

- For small perturbations of the initial nuclide composition, first order perturbation theory yields good results for the change of *BG*. Non-perturbative calculations show that for a 5% perturbation of the most abundant fuel nuclide (^{238}U), FOPT still gives a good result compared to the direct, non-perturbative forward calculation, as indicated in table 5.6.

- The change of the nuclide composition during cool down of the fuel cannot be neglected in the closed fuel cycle. An important effect in this regard is the decay of fissile ^{241}Pu to the absorbing ^{241}Am.

- Inspection of the solution of the adjoint transmutation equations can give an idea as to how changing a certain parameter (irradiation time, cool down interval) influences the resulting nuclide composition, and thus the fuel cycle.

- The closed fuel cycle can be achieved if the reprocessing efficiency is high enough, i.e. losses during reprocessing should not be too large. If there are losses in reprocessing, adding a small amount of MA to the fuel will yield $BG = 0$. This strategy requires fertile material and some MA, and is thus an almost closed fuel cycle.

- For GFR600, the closed fuel cycle cannot be obtained using the reference fuel and a cool down period of 6 years ($BG \approx -20\%$). Substituting 5% of the uranium with MA gives a BG almost zero. To increase BG over the cycle of irradiation, cool down and reprocessing, increasing the fertile fraction in the fuel is a good option (see table 5.6).

6

Passive reactivity control: Lithium Injection Module

The Generation IV GCFR has a high core power density (for a gas cooled system), and low thermal inertia in the core. This leads to the possibility of rapid temperature excursions. To increase safety margins, design measures are taken, e.g. application of dispersed fuel in ceramic cladding (high retention of fission products), high-melting ceramic structures, and low void effect. In all fast reactors, core restructuring due to excessive temperatures may lead to reactivity accidents [Waltar and Reynolds, 1981]. An adequate, highly reliable SCRAM system is therefore necessary. In this chapter passive reactivity control devices are presented for a GCFR. Using fully passive devices rules out the possibility of unprotected transients, i.e. the fission chain reaction is automatically stopped for off-nominal conditions.

6.1 Passive reactivity control: options and constraints

To shut down a nuclear reactor, three options are generally available:

1. Introduce a parasitic absorber

2. Increase leakage from the reactor

3. Remove the fuel, or reshape the fuel into a less reactive configuration (possible in mobile fuel reactors, e.g. molten salt reactor, pebble bed reactor)

Option 1, the introduction of an absorber, is commonly used to control nuclear reactors. The first decision to be made for passive devices is which quantity should provide the activation signal. In a first iteration, it was decided to use the reactor pressure as the governing parameter. This was motivated by the fact that a Loss Of Coolant Accident (LOCA) in a high power core would have severe effects: temperatures in the core will increase due to a lack of cooling, and the void coefficient may be positive for a GCFR. However, pressure-operated devices are not practical for several reasons:

- There will be a pressure holding system on the primary circuit, which will keep pressure more or less constant, also during the beginning of a transient.
- A change of the reactor power level may introduce pressure variations in the primary circuit. Depending on the Power Conversion System, power output may be controlled by flow (direct cycle) or inventory (indirect cycle) adjustments, leading to pressure fluctuations. If the reactor is intended to do load following, the design pressure in the primary system may have a large range.
- Pressure operated devices need to be disarmed in case of an intended depressurization. These intended depressurizations might occur during maintenance and refueling for instance, when it is advantageous to use a lower than nominal operation pressure in the primary system.

All these issues make pressure operated devices not very practical. The risk of a depressurization is quite small, and at the same time there are other events (Possible Initiating Events, PIE) which might lead to reactor damage as well. Therefore it was decided to design passive devices using the core outlet temperature as the governing parameter. The rationale for this choice is the following: an accident is basically an unintended mismatch between power production and power removal from the core, and any accident will thus manifest itself as a change of outlet temperature, assuming that the inlet conditions are not directly affected by the accident. The activation temperature of the device can be tuned by using a material with the required melting point. The activation temperature of a passive device can have some margin over the maximum design temperature during normal operation and intended transients. A temperature controlled device can be designed to not require disarming for refueling etc. For the proposed passive shut down devices to be successfully integrated into the design of the nuclear reactor, the following demands and constraints are taken into account:

- The passive device is intended to limit the reactor power under accidental conditions when all other active control systems (have) fail(ed).
- The device should be small enough to not interfere with other control systems in the reactor, and should preferably be integrated into the regular control assemblies. The assemblies housing passive devices should not be higher than regular assemblies (see figure 6.1 for background).
- Even though the passive device is to be integrated into a control assembly, the temperature sensitive element should be in the outlet gas stream of a fuel assembly to adequately detect the hot gas outlet temperature.

- The reactivity effect of the passive devices should be so large that activation of only one device will introduce a sufficient amount of negative reactivity.

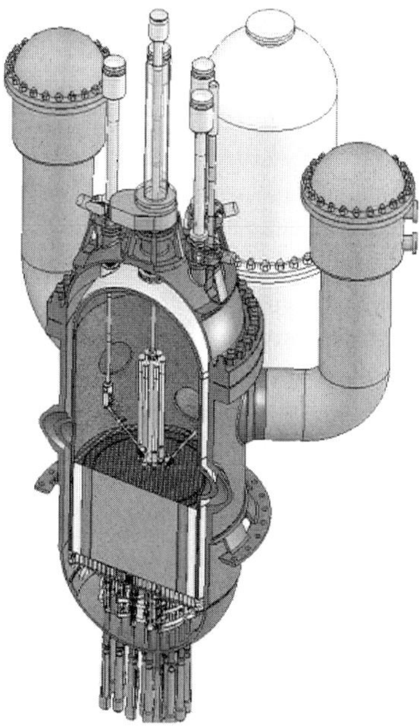

Figure 6.1. *Overview of the core of a Generation IV GCFR (GFR2400, a 2400 MWth Generation IV GCFR design by CEA). The RPV is much higher than the core itself, which is a common feature for all GCFRs (cf. figure 2.2). The reason is the fuel handling: refueling is done under partial pressurization, hence the openings in the RPV to transfer fuel assemblies in and out of the RPV are kept as small as possible. Thus, all fuel assemblies are raised vertically from their diagrid positions, then moved horizontally, and off-loaded through the central opening in the RPV. The space above the core is thus at least as high as the fuel assembly length.*

The last item follows from the behavior of the well known Point Kinetics (PK) equations for the time-dependent neutron flux (or power) in a nuclear reactor [Ott and Neuhold, 1985]. Consider the Constant Decay Source (CDS) approximation of the PK-equations in an initially critical reactor:

$$P_n(t) = P_n(0)\exp(\frac{\rho-\beta}{\Lambda}t) + P_n(0)\frac{\beta}{\beta-\rho}\{1-\exp(\frac{\rho-\beta}{\Lambda}t)\} \tag{6.1}$$

Here $P_n(t)$ is the time dependent neutronic power, ρ is an introduced reactivity, β is the effective delayed neutron fraction, and Λ is the neutron generation time. For $\rho < \beta$, the exponentials on the RHS decay quickly, and the power immediately after the reactivity insertion is approximately given by the Prompt Jump Approximation:

$$P_n(0^+) = \frac{\beta}{\beta - \rho} P_n(0^-) \tag{6.2}$$

with $P_n(0^-)$ the steady state power before the start of the transient, and $P_n(0^+)$ power immediately after introduction of the reactivity. The amplitude of the prompt jump equals:

$$P_n(0^+) - P_n(0^-) = \frac{\rho}{\beta - \rho} P_n(0^-) \tag{6.3}$$

If ρ is large, negative and introduced in a step-wise manner, the power level of the reactor decreases immediately. The magnitude of the prompt jump increases with the magnitude of the reactivity. Once the prompt jump has occurred, the outlet temperature of the coolant will decrease rapidly. If several temperature-controlled passive devices are present, one will be activated, and due to the resulting power decrease the other devices will likely not be activated.

6.2 Neutronic design

In the following the neutronic design of the passive devices is discussed. All calculations presented in this chapter concern the GFR600 (see table 5.1 for design data, table 6.1 for dynamic parameters). The GFR600 core contains 112 fuel assemblies, 6 control assemblies, 3 shutdown assemblies, and is surrounded by 210 reflector assemblies (Zr_3Si_2 reflector). All neutronic calculations were done using the CSAS26 module in SCALE 5 [Oak, 2005] (nuclear data in AMPX-format, 175-group VITAMIN-J group structure). In this model, the fuel assemblies are represented by cell-homogenized mixtures (i.e. self-shielding is taken into account). The reflectors are represented as homogeneous mixtures without any lattice effect, and the absorbers are modeled in their actual geometry, using a cell-calculation to correct the cross sections for effects of absorber self-shielding. Delayed neutron production and decay, and the generation time were calculated using VAREX [Kloosterman and Kuijper, 2000] on the unit cell level (one slab of fuel plus coolant).

To make a temperature controlled passive introduction of reactivity, a device with a freeze seal is chosen. Once the freeze seal melts due to overheating, an absorber is irreversibly released into the core region. To identify the nuclide with the highest negative effect on k_{eff}, a unit cell calculation was done using the sensitivity and uncertainty module TSUNAMI-1D in SCALE 5. The result is given in table 6.2.

Because Eu (natural composition 47.8% Eu-151, 52.2 % Eu-153) and B-10 have the highest effect, the compounds Eu_2O_3 and B_4C were initially selected for further evaluation. Both these compounds are solid at GFR600 temperatures, so they could be introduced by gravity into the core in the form of small spheres or rods. In both cases, the materials are maintained at the hot side of the reactor during steady state, and this may be detrimental to the operation of the device. For example, it is possible that the small spheres fuse together in the high temperature environment, preventing their flow into the core. Therefore, ^{6}Li was finally

Table 6.1. *Dynamic parameters and reactivity coefficients for GFR600. The fuel contains 5% Minor Actinides (see the composition in table 5.2). The Fuel Temperature Coefficient (FTC) is expressed as $\Delta\rho/\Delta T$ [pcm/K]. The Void Coefficient (VC) is calculated as the reactivity difference between 70 bar and 1 bar, and is expressed in pcm per 69 bar. The difference of VC between fresh fuel and the irradiated fuel is contributed to spectral hardening: the spectrum of the irradiated fuel is harder than the fresh spectrum, hence the absence of helium does not affect the spectrum of the irradiated fuel significantly.*

	Fresh fuel	Avg. FIMA 9.33%
β	373.7 pcm	348.2 pcm
Λ	7.5e-7 s	7.5e-7 s
FTC	-0.56 pcm/K	-0.81 pcm/K
VC	+406 pcm	+39.6 pcm

Table 6.2. *Reactivity effects of nuclides in a GFR600 cell calculation using TSUNAMI-1D. The cell calculation includes fuel, cladding and moderator. The absorber is homogeneously mixed into the moderator to obtain the sensitivities. The reactivity effect is expressed in $\Delta\lambda/\Delta N_i$, i.e. the change in reactor eigenvalue as a function of a small change of the number density of isotope i.*

Isotope	$\frac{\Delta\lambda}{\Delta N_i}$
^{151}Eu	-1.40
^{151}Sm[a]	-1.06
^{154}Eu[b]	-0.978
^{10}B	-0.942
^{155}Gd	-0.921
^{153}Eu	-0.827
^{6}Li	-0.366

[a]Unstable fission product, $T_{\frac{1}{2}} = 93$ y.
[b]Unstable fission product, $T_{\frac{1}{2}} = 8.8$ y.

selected because it is liquid at GFR600 operating conditions (see table 6.3). The resulting device is similar to the LIM (Lithium Injection Module) proposed for the RAPID reactor [Kambe and Uotani, 1997]. A preliminary design was made of a LIM: 7 SiC tubes are filled with ^{6}Li in case of off-nominal conditions. A total of 4 LIMs are present, their locations will be discussed later. The storage tank containing the ^{6}Li during normal operation is located in the bottom of the assembly. The location in the bottom is chosen because then the storage tank is in the cold gas stream. In activated mode, the Li should occupy the fueled length of the core (1.95 m). The inner diameter of the pins is chosen as 22 mm, giving a required volume of ^{6}Li of 5200 cm^3 for the 7 pins. The size of the storage tank in the bottom of the assembly should allow for the regular control elements to be operable. The height of the tank is chosen as 45 cm, giving a surface area of some 116 cm^2, which is roughly 50% of the

Table 6.3. *Lithium properties*

T_{melt}	453.69 K (= 180.54°C)
T_{boil}	1615 K (=1341.85°C)
ρ [298 K]	535 kg/m^3
λ	85 W/mK

Table 6.4. *The reactivity effects of the LIMs. Each LIM contains 7 tubes with ^{6}Li, r_{in} = 1.1 cm, r_{out} = 1.2 cm. The LIM worth increases with burnup, which is mainly contributed to the decrease of β: from 373.7 pcm to 348.2 pcm.*

	k_{eff}	ρ	ρ/β	$\Delta\rho/\beta$
	Fresh fuel			
0 LIMs	1.02432 ± 0.00027	0.0237	6.35	-
1 LIM, center	1.01613 ± 0.00023	0.0159	4.25	-2.1
1 LIM, off center	1.01848 ± 0.00026	0.0181	4.86	-1.49
4 LIMs	0.99922 ± 0.00023	-7.8e-4	-0.21	-6.56
	Average burnup 9.33% FIMA			
0 LIMs	1.01373 ± 0.00022	0.0135	3.89	-
1 LIM, center	1.00587 ± 0.00025	0.0058	1.68	-2.21
1 LIM, off center	1.00826 ± 0.00029	0.0082	2.35	-1.54
4 LIMs	0.98868 ± 0.00025	-0.011	-3.29	-7.18

cross flats surface area within the assemblies. The storage tank should be properly shielded from (thermal) neutrons to prevent premature depletion of the lithium. The tank is located as deeply as possible in the assembly, to maintain a good height of reflector and shielding above the tank. The 7 SiC tubes occupy some 13.7% of the volume of the assembly, so enough room is left over for the regular control rods and mechanisms. As seen from table 6.3, lithium is solid at room temperature. It may not be necessary to have an external heating mechanism to keep the lithium fluid at all times. Firstly, The volume of lithium is so small that it will easily melt during the start-up phase of the reactor, if it is solid. It is not expected that the reactor will be taken far below nominal temperatures during refueling to avoid thermal stresses on the RPV and internals. It is likely that the lithium will stay fluid without an external heating mechanism.

The reactivity worth of the LIMs was calculated for several permutations: no LIMs active, one LIM active in the central position of the core, one LIM active in an off-center position, and all four LIMs active. The reactivity worths are given in table 6.4.

The LIM is illustrated in figure 6.2. The 7 SiC tubes are connected to a plenum at the very top of the assembly. From this plenum, tubelets are protruding sideways into the hot outlet gas stream of the neighboring fuel assemblies. These tubelets contain a small freeze seal from a material with the required melting temperature. Candidate materials for the freeze seals could be silver (T_m = 962°C) or gold (T_m = 1064°C, maybe adding some alloying element to lower the melting point). The freeze seal should have a small volume to make sure it responds

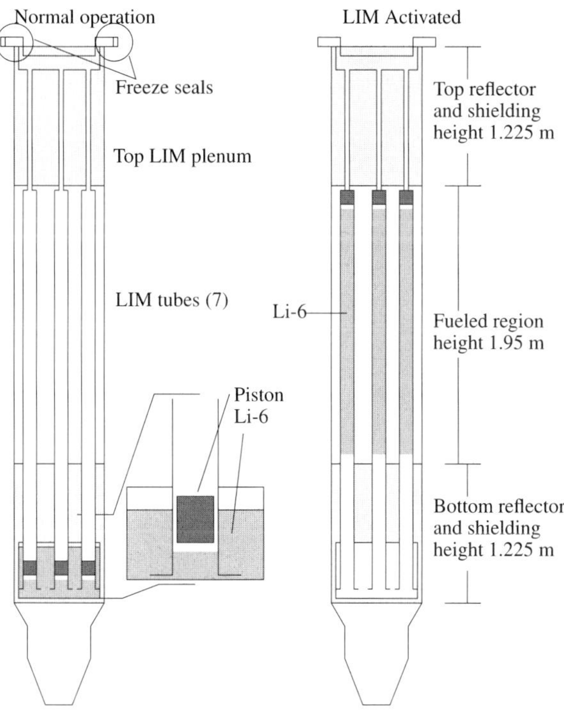

Figure 6.2. *Rough design of a Lithium Injection Module (LIM), activated by a freeze seal (not to scale). The height of the storage tank is some 45 cm. When the LIM is activated, the small SiC pistons seal off the tubes to prevent the ^{6}Li from spraying into the reactor. The pressure in the storage tank is above the reactor operating pressure.*

adequately to a temperature increase. The fabrication of the freeze seal should be as uniform as possible. If one of the freeze seals ruptures, the ^{6}Li shoots up in the tubes, into the fueled region of the core. The storage tank should be adequately pressurized to move the ^{6}Li from the tank into the fueled region of the core. The 7 tubes with ^{6}Li are equipped with small SiC pistons, to seal off the top of the tube in order to prevent the ^{6}Li from spraying into the core after activation. These small sealing pistons should have a 'one-way' operation: once they have moved to the top of the assembly, they should stay there rigidly, to make sure that the ^{6}Li is not pushed back into the storage tank if the reactor pressure increases. A layout of the GFR600 core with the LIM assemblies is shown in figure 6.3. The locations of the LIMs are chosen to obtain the largest number of fuel assemblies connected to a LIM assembly.

In IAEA-TECDOC-626 [1991] four categories of passivity ('A' to 'D') are defined for components and systems of a nuclear reactor. The proposed LIM would fall into Category 'C': *no signal inputs, no external power source or forces; with moving mechanical part(s) and/or moving working fluid; fluid movement due to thermohydraulic conditions.*

From figure 6.2 it is seen that it is advantageous if the column of Li can be suspended into the

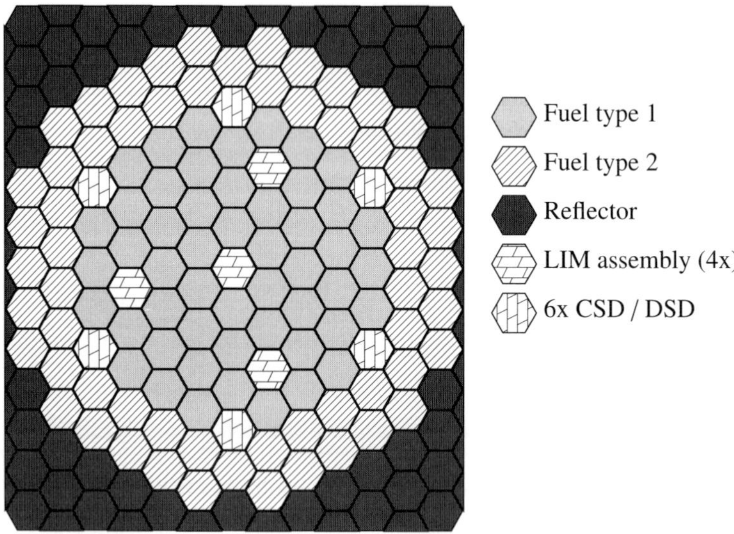

Figure 6.3. *The layout of the GFR600 core with LIMs. The locations of the LIMs are chosen to obtain the largest number of fuel assemblies connected to a LIM.*

fuel zone, because the required volume of ^{6}Li is smaller. A column of fluid can be maintained over a less dense fluid if the diameter of the tube is small enough. The characteristic tube size is given by the limit for the so-called Rayleigh-Taylor instability [Guyon et al., 2001]. An order of magnitude calculation shows that for liquid lithium over helium the characteristic size is about 50 mm. Since an inner diameter of 22 mm has been chosen, it is concluded that the liquid lithium can be suspended.

6.3 Thermalhydraulic model of GFR600

A simplified thermalhydraulic model of the GFR600 core was made based on design data from the GCFR-STREP report on GFR600 [Conti and Bosq, 2004], and on data supplied by CEA [Dumaz et al., 2006]. Refer to figure 6.4 where an overview of the main design features of the primary system is given. The figure is quite generally applicable for all Generation IV GCFRs, all of which have a similar system layout. Cold helium enters through the annular cross duct, flows down through the downcomer into the lower plenum and into the fuel assemblies. The flow distribution of the coolant over the individual assemblies is done with flow gags on each assembly. The upper plenum is a large volume filled with hot helium, with the hot leg of the cross duct connected to it, as well as 3 Decay Heat Removal (DHR) loops. To protect the Reactor Pressure Vessel (RPV) from the hot coolant, the entire primary circuit is laid out as cross ducting, with cold helium flowing between the outer RPV and an inner isolation liner. The RPV thickness is 75 mm, the vessel material is, depending on the operating temperatures and conditions, 9Cr1Mo steel, or stainless steel [Morin et al., 2006, Bailly et al., 1999].

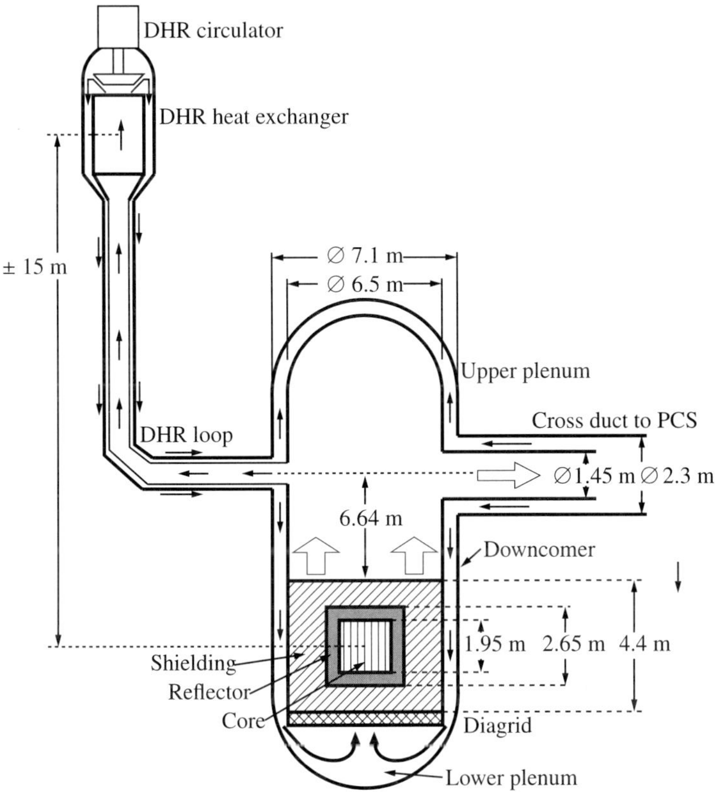

Figure 6.4. *Layout of the primary circuit of GFR600. The RPV would have openings at the top to allow refueling, and at the bottom for the control rod actuators. One of the three redundant DHR circuits is shown. The DHR circulator provides adequate cooling under depressurized conditions. To control the flow of coolant, valves are present. The most marked difference with older GCFR designs is that the RPV is made of steel instead of concrete.*

The DHR strategy is similar for all Generation IV GCFRs. Under pressurized conditions natural circulation of the coolant through the DHR circuits should be sufficient for Decay Heat Removal. To assure adequate natural circulation, the elevation between the core mid-plane and the DHR heat exchanger is some 15 m. The DHR loops are designed to each extract at least 3% of the nominal reactor power. During normal operation measures have to be taken to isolate the DHR-circuits from the primary coolant flow using valves. Thus, according to IAEA-TECDOC-626 [1991], the natural circulation DHR-loops qualify as category 'D': *Execution of the safety function through passive methods; but using an external signal to initiate the process (opening of valves, etc).* Note that the valves should be *'suitably qualified'*, according to the IAEA specifications.

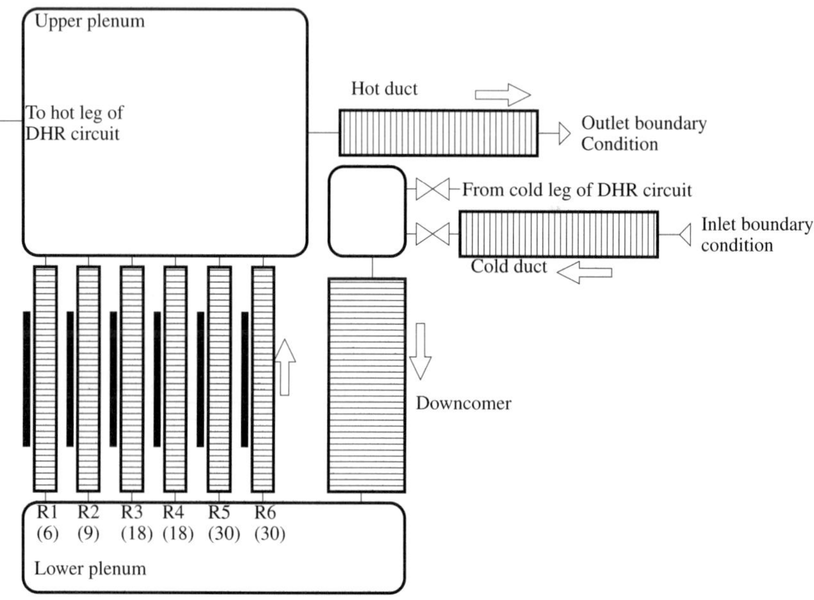

Figure 6.5. *Illustration of the CATHARE model of the GFR600 primary system. R1 through R6 are the 6 rings of fuel assemblies, the number in brackets the number of fuel assemblies per ring. The black structures represent the fuel plates. All cross ducting is represented as concentric pipes, i.e. heat exchange between hot and cold fluid is neglected.*

To obtain an adequate flow in the DHR circuits under depressurized conditions, a small DHR circulator is present. The entire primary system is to be surrounded by a 'close containment', which should maintain a backup pressure of some 10 to 15 bars under all circumstances. The required pumping power for the DHR circulators will then be so small that they can be driven on battery power for 24 hours (according to IAEA-TECDOC-626 [1991] such devices are not rated 'passive').
The primary circuit of 6.4 was modeled using the CATHARE code [Lavialle, 2006]. This code (CATHARE2 v2.5 mod3.1, developed by CEA, EdF, Framatome-ANP and IRSN), is a two-phase implicit transient thermalhydraulics code, able to perform calculations on helium cooled reactors. As a detailed description of the power conversion system does not exist, the inlet and outlet of the reactor are described as boundary conditions . The fuel plates are modeled as flat plates in which power is produced. The CATHARE model (figure 6.5) has 6 fuel elements, each representing one ring of fuel assemblies. The axial power profile has

a cosine shape with a power peaking of 1.3, radial power peaking is 1.15. Bypass flow in between the fuel elements, as well as flow in the reflector, are currently not modeled. The flow gags on the fuel elements are set up to get a uniform outlet temperature of 850°C. Heat exchanges between the vessel and the surrounding environment are not taken into account, due to a lack of detailed design. The secondary fluid of the DHR heat exchangers is water. The secondary fluid is driven by natural circulation. The DHR circuits are isolated during steady state by valves on the cold side. A schematic of the CATHARE model of the entire circuit is given in figure 6.6.

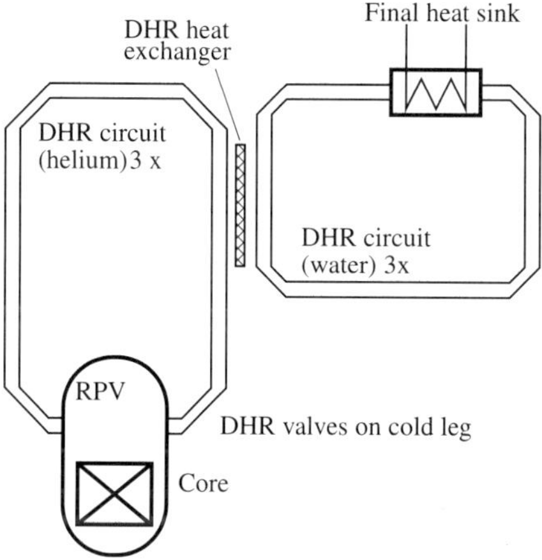

Figure 6.6. *Illustration of the CATHARE model of the GFR600 primary system with DHR loops. The water on the secondary side of the DHR circuit flows under natural circulation. Heat is transferred to a pool of water, which is the final heat sink.*

The LIMs are each connected to their six neighboring fuel assemblies, with the freeze seal protruding into the hot outlet gas stream of the fuel assemblies. From this configuration some important design issues are identified:

- If a localized transient occurs, one LIM will be activated, and the power goes down. The other LIMs will not be activated, unless the transient inserts more reactivity than the worth of one LIM.
- LIMs only respond to events in one of the connected fuel elements. If the LIMs are to detect highly localized events like a flow blockage, each fuel assembly should be connected to at least one LIM.
- The freeze seals have to as uniform as possible in order to activate the largest number of LIMs for a given transient.

Given the simplified primary circuit description, three relevant core-wide transients were calculated:

1. A loss of flow at constant pressure. An initiating event for such a transient would be the (faulty) activation of the turbocompressor braking system (the braking system prevents the turbocompressor from over-speeding, for instance due to a load rejection).

2. Ramp insertion of reactivity, 0.9$ of reactivity is introduced linearly over 20 seconds to simulate the faulty withdrawal of a control rod. Pressure and flow remain at steady state values.

3. Ramp insertion of reactivity, but now 0.9$ in 1 second, to simulate the effect of a control rod ejection.

For all transients the LIMs are activated if the gas temperature at the outlet of (one of) the fuel assemblies surpasses 1000°C. Once the LIM is activated, introduction of the ^{6}Li is simulated to take one second. All transients were calculated assuming that one LIM is activated. If one LIM is activated, the LIM worth depends on the position of the LIM. As given in table 6.4, one LIM at the center position introduces -2.1$, while one LIM at an off-center position has a worth of -1.49$. An extra calculation was done, in which one LIM with a smaller worth (-1.1$) is used. This calculation serves to estimate opportunities to change the LIM design.

Calculation of reactor power during a transient

The total thermal power of a nuclear reactor is due to power released by fissions (fission power), and partly due to the decay of radioactive fission products (the residual or decay power). The decay power is commonly described as the time dependent power following one fission in isotope j as:

$$f(t) = \sum_{k=1}^{K} \gamma_{jk} \exp(-\mu_{jk} t) \tag{6.4}$$

This formulation assumes that all fission products can be grouped into K groups, each decaying with their respective time constant μ_{jk}, with a contribution γ_{jk} to the total decay power. Decay Heat standards based on equation (6.4) include the ANS5.1 standard with $j = 4, k = 23$ [ANSI/ANS-5.1-1994, 2004], the DIN standard ($j = 4, k = 24$, DIN [1990]) or JAERI standard ($j = 5, k = 33$, Tasaka et al. [1990]). The function $f(t)$ is an impulse response and can be extended to an arbitrary fission rate history $\psi(t)$ by convolution. Assuming that the fission rate history extends from $t = 0$ to $t = t_0$, the decay heat at time t_0 is found from:

$$P_d(t_0) = \int_0^{t_0} \psi(t_0 - \tau) f(\tau) d\tau \tag{6.5}$$

The convolution equation (6.5) can be rewritten as a differential equation. Doing so, a coupled set of differential equations is found which describes the time dependent power of a (initially critical) nuclear reactor:

$$\begin{aligned}
\frac{dP_n(t)}{dt} &= \frac{\rho(t)-\beta(t)}{\Lambda(t)}P_n(t)+\frac{1}{\Lambda_0}\sum_{i=1}^{6}\lambda_i\zeta_i(t) && (6.6)\\
\frac{d\zeta_i(t)}{dt} &= \frac{F(t)}{F_0}\beta_i(t)P_n(t)-\lambda_i\zeta_i(t) && (6.7)
\end{aligned}$$

$$\begin{aligned}
\frac{d\epsilon_{jk}(t)}{dt} &= C_c f_j\frac{\gamma_{jk}}{\mu_{jk}}\psi(t)-\mu_{jk}\epsilon_{jk}(t) && (6.8)\\
P_d(t) &= \sum_{j=1}^{J}\sum_{k=1}^{K}\mu_{jk}\epsilon_{jk}(t) && (6.9)
\end{aligned}$$

In this set of equations, $P_n(t)$ is the neutronic power (due to fissions), $\beta_i(t)$ are the effective delayed neutron fractions ($\beta(t) = \sum_{i=1}^{6}\beta_i(t)$), $\Lambda_0, \Lambda(t)$ are the steady state resp. transient neutron generation time, $\zeta_i(t)$ are the precursor concentrations for delayed neutrons, with λ_i the corresponding decay constants, and $F_0, F(t)$ are the steady state resp. transient fission rate. The quantity $\epsilon_{jk}(t)$ is known as the pseudo decay power. C_c is a factor allowing a conservative calculation of the Decay Heat power if set to larger than 1.0, f_j is the fraction of fissions in isotope j, and $P_d(t)$ finally is the resulting decay heat power. If adequate substitutions are made, $\psi(t)$ can be expressed in $P_n(t)$ and a fully coupled set of equations is found.
The model described in equations (6.6) to (6.9) is available in CATHARE2 with 11 decay heat groups. To represent the decay heat source of the GCFR, the decay power is calculated using the ANS5.1 law, employing a fissioning mixture with 20% fast fissions in ^{238}U, 40% fissions in ^{239}Pu and 40% fissions in ^{241}Pu. The curve obtained in this way was fitted with an 11-group model based on equation (6.4). The resulting model gives a steady state decay power of 6.7% of nominal power. In figure 6.7 a comparison is presented of the decay heat power from the ANS5.1 standard for thermal fission of ^{235}U (LWR), the synthetic GCFR curve, and the decay power calculated by CATHARE2 using the fitted decay heat model.
It should be noted that the decay heat curve proposed here might not be very accurate. First of all, no data is available in the ANS5.1 standard for fast reactors. Secondly, alpha decay of the MA is not taken into account. Thirdly, the effect of capture reactions in the actinides and subsequent decay are not treated explicitly. On the other hand, application of the model of equations (6.6) - (6.9) incorporates the entire power history during a transient. The proposed model is the best possible given the currently available data on decay heat, fuel design and material properties.

6.4 Results of transient simulations

Loss of Flow

In all calculations, the coolant mass flow rate is assumed to decrease as

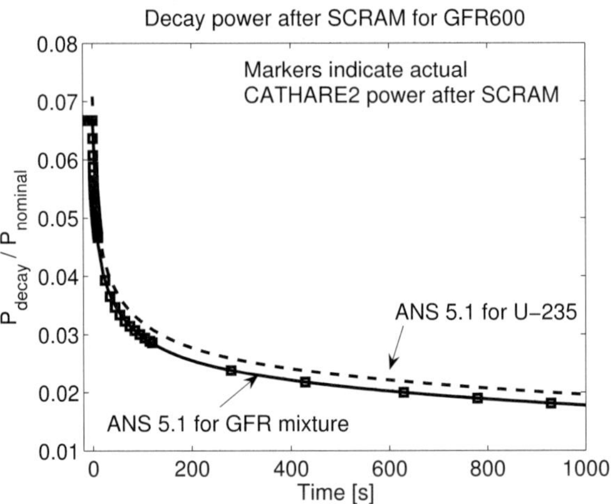

Figure 6.7. *Decay heat curves using the ANS5.1 standard [ANSI/ANS-5.1-1994, 2004]. The solid line is the decay heat curve corresponding to a mixture with 20% fast fissions in* ^{238}U*, 40%* ^{239}Pu*, and 40%* ^{241}Pu*. The rectangular markers indicate the decay heat power after SCRAM calculated by CATHARE2. For comparison, the dashed line gives the decay heat for* ^{235}U*.*

$$\dot{m}(t) = \frac{\dot{m}_0}{1 + t_s/\tau} \tag{6.10}$$

with t_s the time since the start of the transient and $\tau = 10$ s. When the mass flow in the cold duct is decreased below 5% of the steady state value, the cold duct and hot duct are isolated, the valves on the DHR circuits are opened, and natural circulation is allowed to develop.

A first calculation was done for an unprotected LOFA (Loss Of Flow Accident) without any control, to check whether a LIM is necessary. Two calculations were done, one with the Fuel Temperature Coefficient for MA fuel, and one with a ten times stronger temperature feedback. The result is given in figures 6.8 and 6.9. In both cases the reactor stabilizes in the natural circulation regime at a power of some 75 MWth. This power is larger than the DHR loops are designed to extract (18 MWth). In the natural circulation phase, the fuel temperature is some 1200°C, which may be within limits, but such high temperatures are undesirable for a prolonged period. So it is concluded that some reactivity control device is necessary.

Please refer to figures 6.10 and 6.11. When the coolant mass flow rate into the core starts to decrease, the fuel temperature starts to increase, leading to negative reactivity due to Doppler feedback. Once one LIM is activated, the introduction of negative reactivity is so large that the fission chain reaction shuts down quickly. The prompt jump which occurs after activation of the LIM(s) is clearly visible in figure 6.10. Following the prompt jump, the neutronic power decrease is governed by the slowest group of delayed neutron precursors, and by the

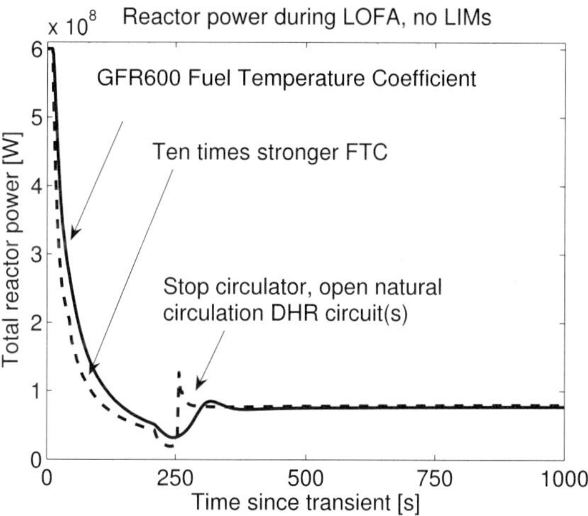

Figure 6.8. *Power during unprotected Loss of Flow. The solid line is for the actual GFR600 Doppler feedback coefficient, the dashed line is for a ten times stronger feedback. In the natural circulation phase, the reactor power stabilizes at 75 MWth, which is more than the DHR circuits are designed for.*

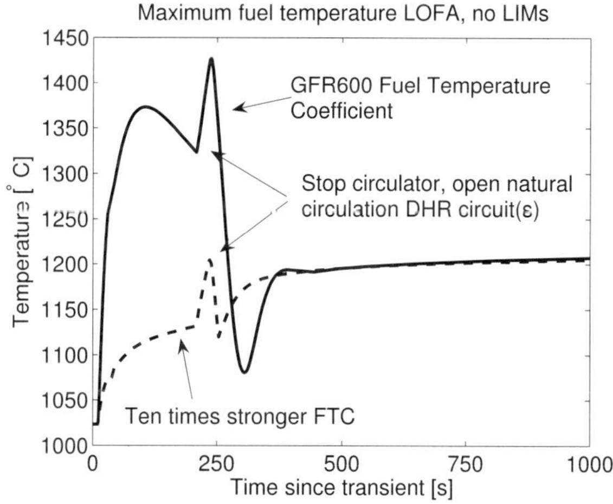

Figure 6.9. *Temperature during unprotected Loss of Flow. The solid line is for the actual GFR600 Doppler feedback coefficient, the dashed line is for a ten times stronger feedback. The reactor stabilizes at 1200°C, which is too high.*

decay power. For a larger worth of the LIMs, the initial jump down is larger (see equation

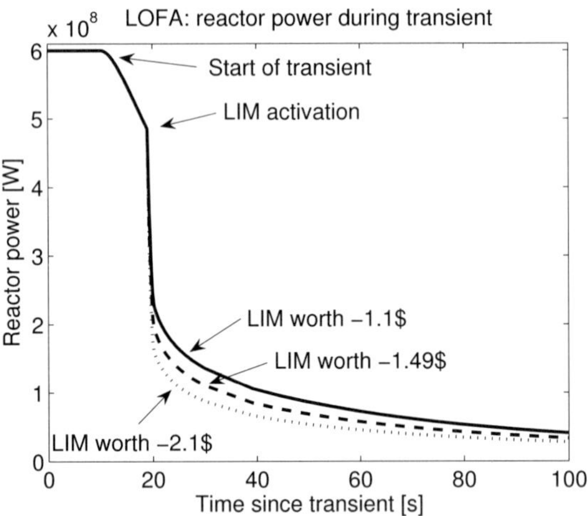

Figure 6.10. *Normalized power during Loss of Flow with LIMs. Note the difference in power just after activation of the LIMs: introducing more negative reactivity reduces the power faster.*

(6.3)). This results in lower temperatures during the rest of the transient. When the cross duct is isolated, and the DHR circuit is opened, a 'hump' is visible in the graph (around 200 s). The hump is caused by a short period of flow stagnation when the cold and hot duct valves are closed. Soon thereafter, natural circulation is established and temperatures in the core start to decrease. Under pressurized conditions, the DHR circuit has adequate heat removal capacity: the outlet temperature of the core decreases steadily under natural circulation.

Reactivity evolution after Loss Of Flow

While the core cools down, its reactivity increases due to the Doppler effect. For GFR600, the reference fuel temperature used to calculate the magnitude of the Doppler effect is 858°C (steady state value). There are three plausible values of the final temperature of the core after SCRAM:

1. 480°C, being the steady state inlet coolant temperature. This introduces +212 pcm (0.57 $) of reactivity.

2. 90°C, being the temperature of the water in the secondary branch of the DHR circuits. This introduces +430.1 pcm (1.15 $) of reactivity.

3. 25°C, which is just ambient temperature. This introduces +466.5 (1.25 $) of reactivity.

These values are calculated based on β of the GFR600 fuel with 5% MA. The Doppler effect is stronger for the standard MOx fuel, leading to larger reactivity effects caused by the cooling

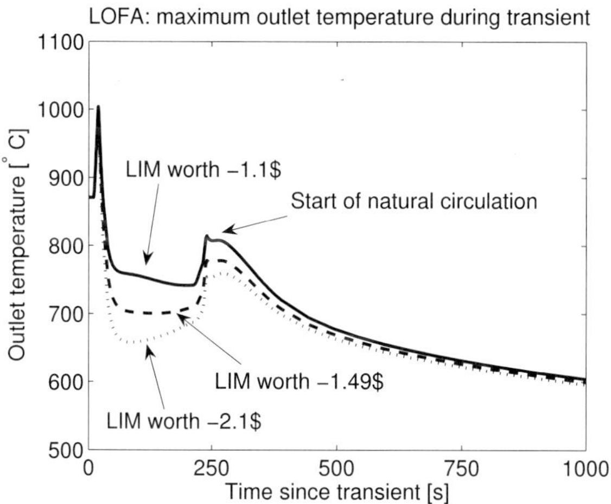

Figure 6.11. *Helium outlet temperature during Loss of Flow with LIMs. The hump in the graph is caused by the isolation of the turbine, followed by the opening of the DHR system. Note the different time scale vs. figure 6.10.*

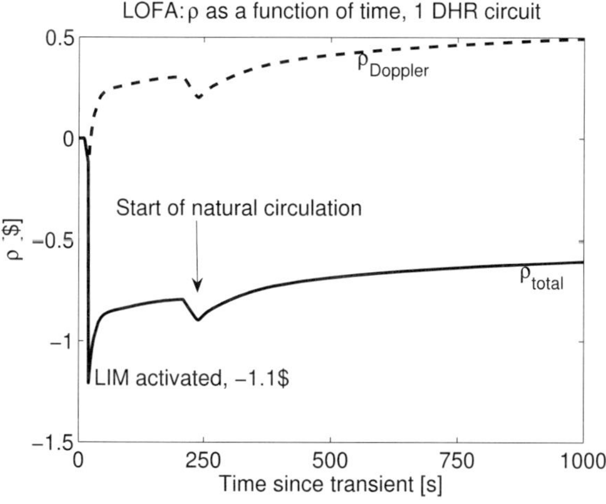

Figure 6.12. *Contributions to the total reactivity during the Loss Of Flow transient, using a LIM worth of -1.1 \$, and only one DHR loop. The Doppler-effect is positive due to the decreasing temperatures, and the total reactivity is negative during the transient.*

down of the fuel. In figure 6.12 the Doppler reactivity and total reactivity are given for a LIM worth of -1.1 $.

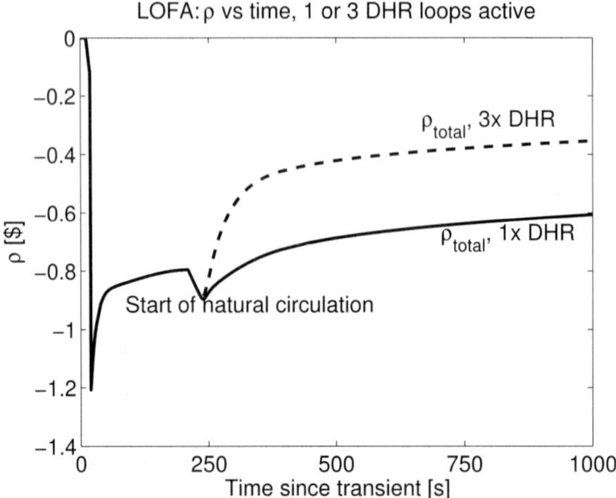

Figure 6.13. *Contributions to the total reactivity during the Loss Of Flow transient, using a LIM worth of -1.1 $. If 3 DHR loops are opened instead of one, the core cools down quicker, and the reactivity margin to re-criticality is smaller.*

The primary system is equipped with 3 redundant DHR loops, each capable of extracting 3% of nominal power under pressurized conditions. If all 3 loops are activated, the primary system will cool down much faster than when only one loop is activated. As a result, reactivity effects due to Doppler feedback will have a different time-scale. In figure 6.13 an illustration is given of this effect. If one DHR loop is activated, the rate at which the core cools down is such that after 1000 s the reactivity is about -0.6 $, but with all three DHR loops active, the reactivity is smaller than -0.4 $ at 1000 s. In the calculations presented here, the temperature of the ultimate heat sink is 90°C, so with a LIM worth of -1.1 $ the core will become critical again. The moment of re-criticality is determined by the number of active DHR loops.

Ramp reactivity introduction

For this scenario, the results are summarized in figures 6.14 and 6.15. In all cases, one LIM would be sufficient to limit the reactor power. If one LIM is activated, a certain amount of negative reactivity is introduced, and together with the +0.9$ due to CR movement and the Doppler feedback, a net reactivity is established. During this transient it is assumed that the coolant inlet conditions remain at steady state values. Hence, the lowest temperature the fuel can attain is 480°C. As shown in the previous section, cooling down the fuel to 480°C increases reactivity by 0.57$. It is concluded that a LIM worth larger than 0.9$ + 0.57$ will shut down the reactor. A lower LIM worth will establish a new power level in the reactor, where the Doppler feedback, LIM worth and CR worth determine the final state.

A calculation was done simulating a control rod ejection, where the CR moves out of the

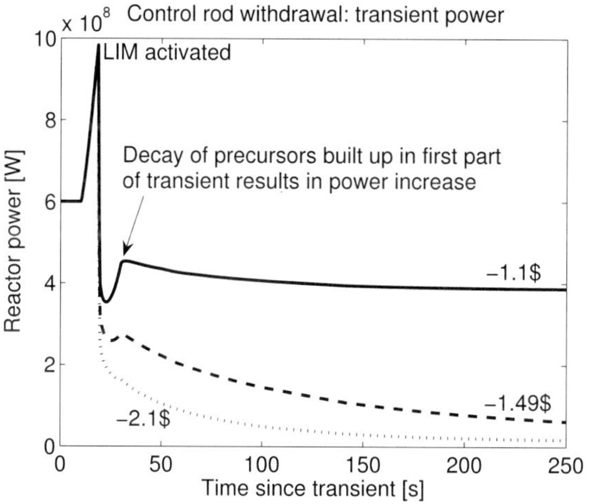

Figure 6.14. *Power during the withdrawal of a control rod (+0.9$) in 20 seconds. The power reaches a maximum of 1.65 times nominal power. After activation of the LIM, the prompt jump results in a quick decrease of power.*

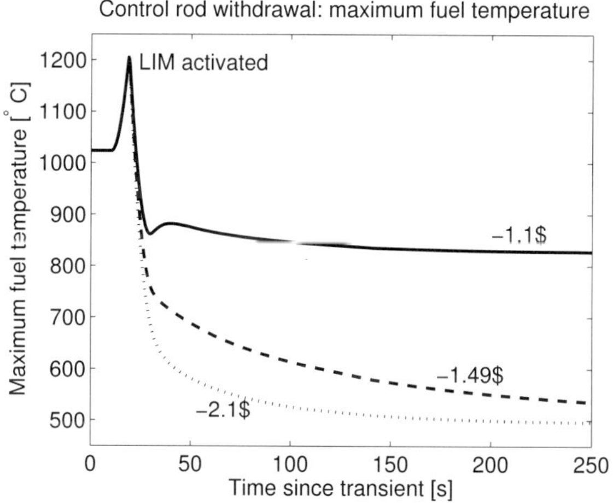

Figure 6.15. *Maximum fuel temperature during the control rod withdrawal in 20 seconds. The peak fuel temperature is some 1200°C. If LIM worth is -1.1$, a new steady state temperature is established which is below the normal steady state temperature.*

core in 1 second. In figures 6.16 and 6.17 the results are given for the reactor power and the maximum fuel temperature. Note that the power peaks at almost 6 times nominal power. The maximum temperature of the fuel is not much higher than in the previous case, about 1350°C.

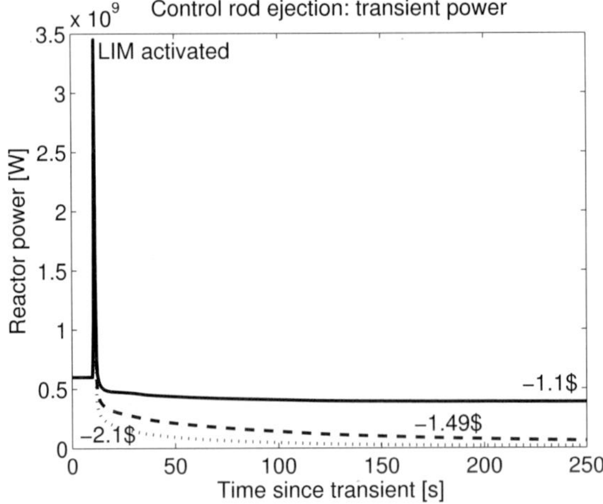

Figure 6.16. *Power after control rod ejection (+0.9$ in 1 second). Note that the power peaks at almost six times the nominal value.*

From figure 6.16 the long-term behavior of the reactor after the control rod ejection is not much different from the control rod withdrawal case, but the beginning of the transient is much steeper, with a very rapid increase of power production and temperatures. This causes concern for material degradation by thermal shock (see appendix B). It should be noted that if an active SCRAM system is used, the resulting transient would be similar to the result given here.

6.5 Conclusions

In this chapter a passive device is proposed to limit the reactor power output under off-nominal conditions when other control elements have failed. The developments in this chapter allow the following conclusions:

- LIMs are capable to control the power production in the reactor, even if a large reactivity is accidentally inserted (+0.9$). If only one LIM is activated, the reactor may not be shutdown completely, but the power production remains bounded.
- The precise requirement for the worth of one LIM cannot yet be determined. The required worth depends on several design parameters, which are not yet known: the worth of one Control Rod, the pressure holding system and the lowest fuel temperature reached in a transient. However, from the calculations presented in this chapter, it is clear that even a rather low worth (-1.1 $) will protect the fuel from excessive temperatures.

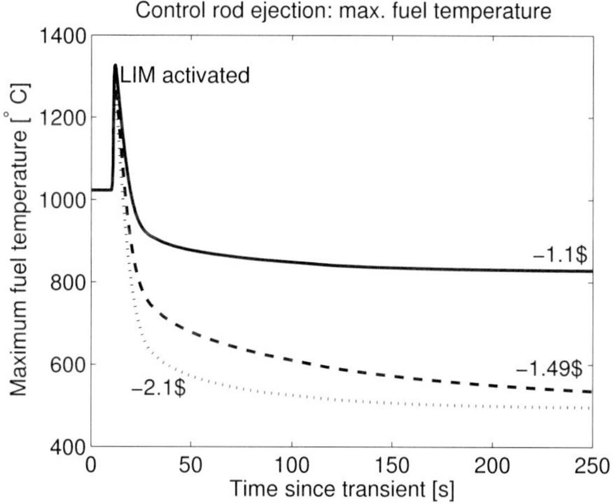

Figure 6.17. *Maximum fuel temperature after control rod ejection. The fuel temperatures change much quicker than in the case of the gradual control rod withdrawal. Such rapid temperature transients may cause material degradation.*

- If a temperature controlled passive control element is present in the core, it is highly likely that only one will be activated. The prompt jump in power caused by LIM operation will decrease temperatures below the activation temperature of a LIM. Redundancy may call for more than one LIM in the core.
- In a gas cooled core, a temperature controlled passive device will only detect off-nominal conditions occurring in a neighboring fuel assembly.
- Rapid transients initiated by (spurious) control rod movement, followed by a quick response of the active or passive control system result in short transients of power production and temperatures. Degradation of ceramic structural materials might become problematic in these cases.

The presence of the LIMs in the core may introduce new (extra) modes of failure. An example could be the spurious activation of a LIM, or the breach of one of the LIM tubes. The risks of the presence of the LIMs are currently difficult to evaluate, as this requires a detailed design of the LIMs. Also, a list of Probable Initiating Events for LIM failure should be prepared if a more detailed LIM design is available.

7
Conclusions and discussion

The key aspect for the Generation IV GCFR is the closed fuel cycle, i.e. a fuel cycle where enough fissile material is bred in the core to allow refueling of the same reactor only adding fertile material. All uranium and other actinides are recycled in the reactor, and only Fission Products are discharged to a repository. The work in this thesis shows that it is possible to obtain the closed fuel cycle with a Gas Cooled Fast Reactor. In fact, it is possible to close the fuel cycle, and transmute some extra MA in the GCFR to reduce existing stockpiles of Np, Am and Cm. This thesis shows that there is a rather narrow band of isotopic compositions which will allow a closed fuel cycle. To obtain a closed fuel cycle with ^{238}U as fertile isotope, the fissile ('driver') fraction of the fuel should be between 16% to 20%.

From the investigations into an 'upscaled' HTR, it was found that a pebble bed type core is not applicable for GCFR if a power density is chosen in the range of 50 to 100 MW/m^3 (preferred range in Generation IV). Coated particle fuel can be used, if the particles are cooled directly (i.e. particles not encapsulated in fuel elements). The coated particle GCFR has a relatively low volume fraction of fuel, and would preferentially be a large reactor, i.e. 2400 MWth at a power density of 50 MW/m^3 as proposed in this thesis.
For safety reasons, the volumetric power density of Generation IV GCFRs is relatively low for a fast reactor. This results in a low specific power, leading to long irradiation intervals and possibly an economic penalty from poor fuel economy. The coated particle GCFR concepts proposed in this thesis have a specific power between 21 and 28 W/gHM, below the tentative range for Generation IV (50 W/gHM). The objective of self-breeding without blankets causes the reactivity swing during irradiation to be low, and possibly positive. It is possible to design for long irradiation intervals, reaching high burnup with a low overreactivity of the fresh fuel.

In this thesis, a new, integral definition of Breeding Gain (*BG*) is proposed. The reactivity weights needed to calculate *BG*, can be defined using an expression based on eigenvalue perturbation theory. For the closed fuel cycle, *BG* should be evaluated from Beginning Of Cycle (BOC) of one irradiation cycle to BOC of the next cycle. If the reactivity weights are taken as constants, first order perturbation theory can be effectively applied to calculate the change of *BG* caused by variations of the initial fuel composition.

The application of this theoretical framework to a GCFR shows that the change of the fuel composition during the cool down period, and the losses during reprocessing can not be neglected in the closed fuel cycle. The most important effect during cool down is the decay of ^{241}Pu (fissile) to ^{241}Am (absorber). As a result, the *BG* during irradiation should be larger than zero, i.e. there should be a net production of fissile material. Adding some MA (5%) homogeneously into the fuel has a beneficial effect on the fuel cycle: transmutation of the MA to fissile isotopes yields a *BG* close to zero from BOC to BOC.

The GCFR safety case requires special attention. The GCFR core power density is high for a gas cooled system. Decay Heat Removal by radiation and conduction alone is not sufficient to cool the core under accidental conditions, especially under depressurized conditions. The currently proposed solution with valve-controlled natural circulation DHR loops is not a 'passive' solution according to the definitions of the IAEA.

In this thesis passive devices are presented to limit power production of the core under accidental conditions when all (active) control systems fail. Such a passive device precludes any unprotected transients. The reactivity effect of the passive devices does not need to be very large to obtain an adequate control of the reactor power during transients without other control systems. The passive devices will not shut down the reactor under all conditions, but the power production remains bounded and temperatures will not be so high as to cause (large) damage.

If temperature controlled passive devices are used, it is highly likely that only one will be activated in case of an accident. Only temperature transients occurring in the assembly where a temperature sensitive element is placed, are picked up. Hence, temperature controlled devices will pick up core-wide transients. Highly localized events, such as a flow blockage, will only be picked up if it occurs in an assembly with a temperature sensitive element in it.

The final conclusion of this thesis is that the fuel cycle can be closed using a GCFR with in-core breeding of fuel. Because the fissile fraction is limited, dense fuels (carbide or nitride) are required. The actual shape of the fuel elements is limited by thermal and mechanical constraints rather than neutronic limits. The GCFR concept is different from thermal HTR reactors, and as such does not inherit the inherent safety features of thermal HTRs. Although the Generation IV GCFR is designed with a focus on safety, passive cooling of the core under all conditions is impossible if a reasonable power density is chosen. The passive devices proposed in this thesis make unprotected transients impossible, but evacuation of the Decay Heat under all conditions remains to be proven.

Discussion and future outlook

Many opportunities remain for future investigations into the GCFR and the closed fuel cycle. Subjects that have not been addressed in this thesis include for instance helium gas production due to α-decay in fuel loaded with Minor Actinides, and the effect of nuclear data uncertainties on nuclide evolution and repercussions on Breeding Gain. From a more theoretic point of view, the special heterogeneity of helium cooled lattices needs attention. The lattice of absorbing fuel lumps and helium gives rise to special neutron streaming paths that require special treatment. The Dancoff-factor for fuel plates at an angle should be derived. A more detailed analysis should be made of the causes and effects of coolant voiding for helium cooled fast reactors, as the situation seems to be more complicated than for sodium cooled fast reactors. In the closed fuel cycle, many actinide isotopes are present in the fuel. Nuclear data for many of these isotope is not known accurately enough. An analysis could be made of the influence of the nuclear data on criticality, void effect, dynamic parameters, etc.

When the research for this thesis was started in late 2002, the Generation IV initiative had just arrived and different GCFR concepts were proposed by various groups around the world. Our research initially targeted a pebble bed type of reactor using a TRISO-based fuel because this fuel type seems well-suited for high temperature operation. In due course it was realized, by us and by others investigating the GCFR concept, that the coated particle GCFRs would not meet the Generation IV criteria in terms of economics, fuel inventory, safety behavior etc. From early 2005 on, the European GCFR STREP effort focused mainly on a 600 MWth reactor (GFR600) with an ambitious plate fuel assembly using matrix fuel. To obtain criticality with the enrichment required for self-breeding, the matrix fuel needs to have an extreme density of fuel. To reduce neutron leakage, the optimal H/D-ratio was chosen, leading to a positive void coefficient. At the time of writing of this thesis (mid-2006), the focus of research is shifting from the 600 MWth (modular) system to a 2400 MWth design. With the same power density as the 600 MWth reactor, the 2400 MWth core is much larger, giving better margins on the neutronics. This reactor requires a less extreme fuel design, and can have a flattened core to reduce the void effect.

High-temperature operation results in material problems. It is for this reason that supercritical CO_2 (S-CO_2)is attracting more and more interest as coolant. S-CO_2 can be used in both direct and indirect cycles. Application of S-CO_2 at a medium temperature (650°C) gives a thermodynamic efficiency close to that of helium at high temperature [Kato et al., 2004]. The difference is caused by the properties of S-CO_2: at the critical point, the compression work nearly vanishes, so even though the turbine produces less work, a larger fraction is available for electricity generation. As a side effect, the turbo-machinery would be much smaller for S-CO_2. The disadvantage of S-CO_2 is the high pressure (25 MPa, 250 bars and up), potentially resulting in a large void effect. Since the properties of supercritical fluids vary strongly around the critical point, dynamic behavior of an S-CO_2 cooled GCFR could be a problem, and the potential for natural circulation Decay Heat Removal should be examined.

Delft, October 2006.

This thesis was typeset with LaTeX on Gentoo Linux. PostScript translation by dvips, and PDF conversion with ps2pdf, using PDFX and /prepress settings.

The LOWFAT code

The transmutation equation for a given nuclide i is given by:

$$\frac{dN_i}{dt} = -N_i(\sigma_{a,i}\phi + \lambda_{x,i}) + \sum_j y_{j\to i}N_j\sigma_{f,j}\phi + \ldots$$
$$\ldots + \sum_k N_k\sigma_{c,k\to i}\phi + \sum_l \lambda_{x,l\to i}N_l + Q_i \quad \text{(A.1)}$$

The transmutation equations for a mixture of nuclides consist of a set of coupled ODEs, which can be written in matrix notation as:

$$\frac{\partial\vec{N}}{\partial t} = \underline{M}\vec{N} + \vec{Q} \quad \text{(A.2)}$$

The transmutation matrix $\underline{M}$ contains all nuclear data, i.e. σ_x and λ_x, the yield data y_x, and the flux ϕ. The adjoint nuclide equations are given by [Ott and Neuhold, 1985, Williams, 1986]:

$$\frac{-\partial\vec{N}^*}{\partial t} = \underline{M}^*\vec{N}^* + \vec{Q}^* \quad \text{(A.3)}$$

Since the matrix $\underline{M}$ contains only real numbers, its adjoint is given by the transposed matrix, i.e. $\underline{M}^* = \underline{M}^T$. Thus, in principal, if a code is capable of solving the forward equation (A.2), that code can also solve the adjoint equation (A.3). Since no adjoint capable code was available at the time of research for the European GCFR STREP (FP6), it was decided to develop a code capable of solving both the forward and adjoint transmutation equations. Since this code is based on the ORIGEN-S formalism [Gauld et al., 2005], it is called LOWFAT: *L*ike *O*rigen *W*ith *F*orward and *A*djoint *T*ransmutation.

A.1 Obtaining the nuclear data for the transmutation matrix

Solving the transmutation equations through the matrix formulation is equal to the implementation of the ORIGEN-S code. In the SCALE system, the module COUPLE produces a nuclear data library for ORIGEN-S, which contains, amongst others, the complete transmutation matrix $\underline{M}$. It was decided to read the ORIGEN-S data library produced by COUPLE in LOWFAT. In this way, the data can be tailor-made for a specific problem by prepending a cell calculation (CSAS). In LOWFAT, the data is read from the library, and the total matrix $\underline{M}$ is expanded (in ORIGEN-S only non-zero elements of $\underline{M}$ are kept in memory, but modern computers can easily store all data in memory). The current version of LOWFAT only calculates the evolution of actinides, so only the actinide data are used in the actual calculation. Fission products can be calculated as well in LOWFAT. Adding the fission products increases the size of $\underline{M}$, from roughly 100 x 100 to roughly 2000 x 2000. The decay constants of the fission products also significantly influence the calculational effort, and thus computer time, to reach the solution. Therefore, only actinides are treated in the current version of LOWFAT.

A.2 The stiffness problem

The transmutation equations are an example of a so-called stiff system. Stiff means that there are time scales in the matrix $\underline{M}$, that are (much) shorter than the time step Δt used in the numerical integration. For instance, there may be decay constants λ corresponding to a half-life of seconds, whereas a typical irradiation history equals a hundred days. One solution is to choose Δt short enough to allow straightforward integration, but the required number of time steps (and thus computing time) is enormous. The solution for LOWFAT is to use a stiff solver: DVODE [Brown et al., 1989]. DVODE automatically determines the solution type required for the problem at hand. For example, the optimal integration time step Δt is determined automatically for each integration step. In fact, in LOWFAT only the final time needs to be specified, and DVODE will determine the optimal time discretization to integrate between the start time and the end time.

A.3 Comparison between ORIGEN-S and LOWFAT

The forward and adjoint transmutation equations form a pair, both dependent on $\underline{M}$. To obtain the most consistent solution, the data in $\underline{M}$ should be identical between forward and adjoint cases. More specifically: the flux ϕ must be constant over the entire time span covered in the forward irradiation calculation.

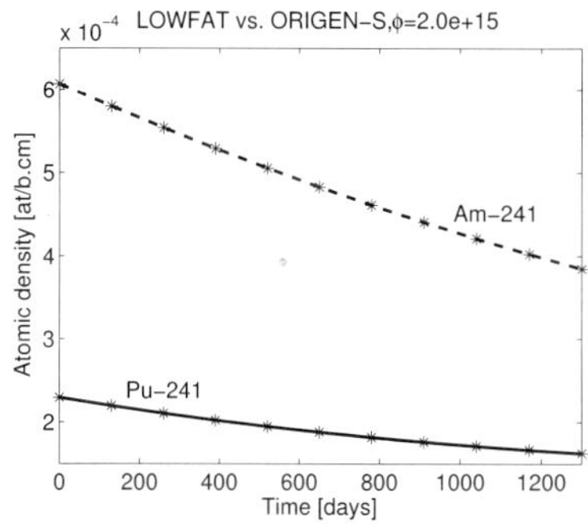

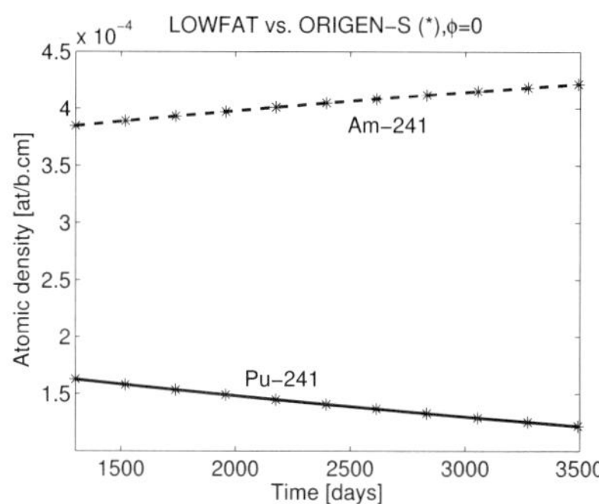

Figure A.1. *Left side: Atomic densities of ^{241}Pu and ^{241}Am during irradiation in a GCFR. Right side: Densities of ^{241}Pu and ^{241}Am during cool down. The solid lines are calculated by LOWFAT, the $*$ indicate the ORIGEN-S results, which are virtually identical.*

To check the accuracy of LOWFAT vs ORIGEN-S, a simple calculation was done on a representative GCFR fuel mixture, with a 1300 day irradiation at constant flux $\phi = 2.0 \cdot 10^{15}$ [n/cm^2.s], followed by a 2192 day cool down period. Results were compared to the same calculation with ORIGEN-S. As illustrative examples, ^{241}Pu and ^{241}Am were chosen, because the half life of ^{241}Pu (14.4 years) is comparable to the calculated cool down period of 6 years. ^{241}Am is the daughter product. In figure A.1 the nuclide densities during irradiation and cool down are given. Solid lines indicate the result of LOWFAT, the results of ORIGEN-S are indicated with $*$. The results for LOWFAT and ORIGEN-S are virtually identical. It was concluded that the results of LOWFAT are adequate.

A.4 Adjoint transmutation calculations

An illustration of the adjoint capability of LOWFAT is given in figure A.2, where several adjoints of ^{238}Pu are illustrated. The adjoint for ^{238}Pu has to be interpreted as follows: the contribution of nuclides present during irradiation to the presence of ^{238}Pu at end of irradiation. An explanation of the behavior of the curves follows:

- ^{238}Pu is itself a source for ^{238}Pu at the end of irradiation. ^{238}Pu nuclei added at the beginning of the irradiation have a chance of transmuting during irradiation, while ^{238}Pu added at the end of the irradiation will not transmute. The curve thus starts low and rises monotonically towards unity at the end of the interval.

- ^{238}Np has a short half life (2.117 days) and decays to ^{238}Pu. Thus, adding a ^{238}Np nucleus has almost the same effect as adding ^{238}Pu, since the half life of ^{238}Np is short compared to the length of the irradiation interval. The curves are close to each other. ^{238}Np nuclei that are added at the end of the interval do not have time to decay to ^{238}Pu, causing the curve to bend sharply downward at the end of the irradiation.

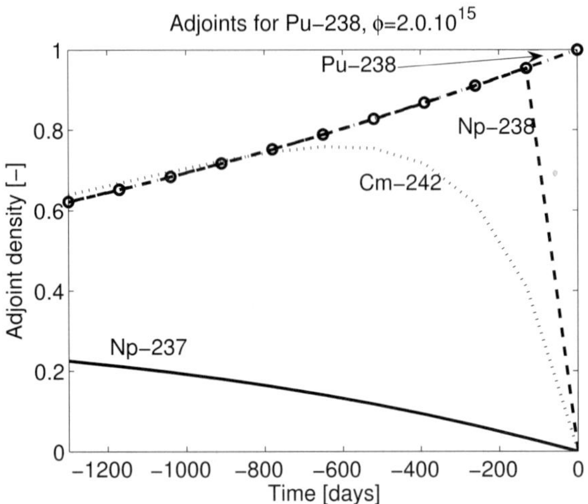

Figure A.2. *Several adjoints for ^{238}Pu, i.e. source isotopes giving rise to the presence of ^{238}Pu at the end of the irradiation interval. ^{237}Np and ^{238}Np contribute by capture and subsequent decay, while ^{242}Cm contributes by α-decay.*

- ^{237}Np contributes by capturing a neutron to become ^{238}Np, and subsequent decay. The difference between the ^{237}Np and ^{238}Np curves is dictated by the capture rate in ^{237}Np. If ^{237}Np nuclei are added too late in the interval, they do not have time to transmute and decay to ^{238}Pu, hence the decreasing trend of the curve towards the end of the interval.
- ^{242}Cm is an illustration of a cyclic production chain: Cm is itself a product of subsequent neutron captures from uranium. ^{242}Cm decays to ^{238}Pu (α-decay) with a half life of 162.8 days. If a ^{242}Cm nucleus is added at the start of the irradiation interval, it will almost certainly decay to ^{238}Pu, and thus ^{242}Cm contributes to the final density of ^{238}Pu in a similar way as ^{238}Np and ^{238}Pu. In fact, ^{242}Cm has a higher adjoint at the start of the irradiation interval than ^{238}Pu itself. This is due to the 'delay' in production. Therefore, the ^{238}Pu produced from ^{242}Cm-decay has a lower chance of transmutation than the ^{238}Pu that is present from the start.

B Properties of ceramics

The structural materials most commonly used in nuclear reactors are zircalloy (cladding in LWRs), stainless steel (structural material in LWRs, cladding in LMFBRs), and graphite for gas cooled reactors (AGR/HTR). The choice for a specific structural material depends on various parameters, like operating temperature and pressure, chemical compatibility with the coolant, behavior under neutron irradiation, machinability etc. As steel looses its mechanical strength above 700 °C, existing HTRs rely on graphite, both as moderator and structural material. Graphite has good high temperature resistance, does not melt in an inert atmosphere (decomposes from 3600°C), has a high thermal conductivity and is an excellent neutron moderator. Graphite is not very strong, and large quantities are needed to obtain adequate strength of graphitic structural elements. In GCFR cores the amount of moderating structural materials should be kept to a minimum, disqualifying graphite as a structural material in GCFRs. The target of high temperature operation, and the impossibility of using graphite determine the selection of ceramics as the structural material in GCFRs. Ceramics are commonly used in high temperature engineering applications. In the following, some basic properties of ceramic materials are discussed.

B.1 Definition and basic properties of ceramics

Ceramics materials are composed of a mix of (semi)-metal atoms and non-metal atoms. It is a broad class of materials, incorporating many well known materials such as sand (SiO_2), salt (NaCl) and SiC. The atomic bonds in ceramics are ionic, covalent or a mix of covalent and ionic. These bonds are generally strong, leading to a high melting point and low plasticity ('strong and brittle materials'). Some ceramics are chemically inert, others are readily soluble in (in)organic solvents such as water. Ceramics for engineering applications are usually categorized into three groups:

1. Oxides: Al_2O_3 (spark plugs for combustion engines, electrical fuses), SiO_2 (quartz), Y_2O_3, UO_2 (nuclear fuel).

2. Nitrides: Si_3N_4 (turbochargers, parts for combustion engines), TiN (wear-resistant coating), AlN (heat sinks in electronic devices).

3. Carbides: SiC, TiC, B_4C.

Ceramics usually have a high melting point, and sometimes decompose before they melt, so they cannot be cast or poured into shape as metals are. To form ceramic objects, individual grains of the ceramic material have to make contact, allowing the atoms to diffuse from one grain into the other, effectively joining the particles together to form one entity. Powders are commonly used, to obtain a large diffusing surface. Sometimes a solvent is added to the powder to make it easier to make a shape (like when using clay: the clay is formed into shape, and then heated in an oven, to remove the water and to join the particles together to form a sturdy object). The diffusion of ceramic atoms is difficult as it is basically self-diffusion with a low driving concentration gradient. Sometimes this is improved by adding a small amount of an extra material, which will facilitate the diffusion of the ceramic atoms. However, care must be taken that this extra phase does not compromise the structure and strength of the resulting ceramic. The best way to get the ceramic particles to bond together is to make a very fine powder, form it into shape and then apply high pressure and temperature over a long period of time to enhance the diffusion process (this is most commonly known as HIPping: Hot Isostatic Pressing, and is applicable for larger ceramic objects). Because of the production process from powder to object, there will always be small cracks and empty spaces in the specimen, where, due to a lack of physical contact, the particles have not diffused into each other. Sometimes this porosity is actually required, for instance in nuclear fuel to facilitate the release of fission gas. As will be shown, the ab-initio cracks effectively determine the mechanical behavior of ceramics.

Oxide ceramics generally behave as thermal insulators, making them unfavorable in high temperature applications. Nitride ceramics are in common use in various high temperature industrial applications. If nitride ceramics are to be used in a nuclear reactor, the nuclear properties of nitrogen have to be taken into account. The nuclear cross section for the reaction $^{14}N \longrightarrow [(n,p)] \longrightarrow {}^{14}C$ is quite high, as illustrated in figure B.1. This has a negative impact on core neutronics and on the fuel cycle, as ^{14}C is a long lived isotope with high radiotoxicity. Production of ^{14}C can be avoided by enriching the nitrogen in ^{15}N (natural abundance 0.37 %), but this does not seem to be viable for bulk amounts as required for structural materials, making nitride ceramics unpractical as structural materials. This effectively leaves the carbide ceramics as the preferred structural material for GCFR cores. SiC is a well known high temperature engineering material and has been selected as the reference material for GCFR application in the Generation IV program.

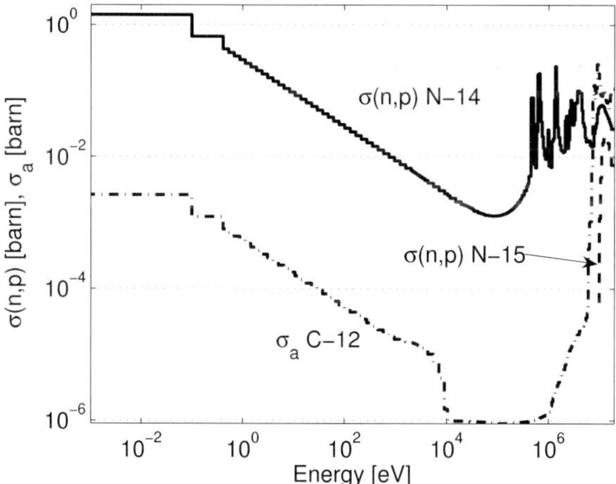

Figure B.1. *σ(n,p) for ^{14}N and ^{15}N. For comparison, σ_a of ^{12}C is also shown. Even though σ(n,p) of ^{14}N is not large on an absolute scale, it is about three orders of magnitude larger than the total absorption cross section of carbon. In the cross section library used to make this figure, the (n,p) cross section for ^{15}N is very small and set to zero for energies below 9.5 MeV.*

B.2 Fracture behavior of ceramics

In this section an introduction is given of the fracture behavior of ceramic materials. It will be shown that the strength of ceramics depends on the (inevitable) presence of small cracks. Consider a group of atoms in a crystal lattice. The forces between the atoms are described by the Lennard-Jones potential [Roest, 1987]:

$$U(r) = -\frac{A}{r^n} + \frac{B}{r^m} \tag{B.1}$$

The interatomic force is then given by:

$$F(r) = -\frac{dU(r)}{r} = -\frac{nA}{r^{n+1}} + \frac{mB}{r^{m+1}} \tag{B.2}$$

The average separation between the atoms can be found by requiring that there is no net force between the atoms, giving:

$$F(r_0) = 0 \rightarrow r_0 = \left(\frac{mB}{nA}\right)^{\frac{1}{m-n}} \tag{B.3}$$

$U(r)$ and $F(r)$ are illustrated in figure B.2. The function $F(r)$ has a maximum for a certain critical separation r_c. If the interatomic bond is extended beyond r_c, the force tying the

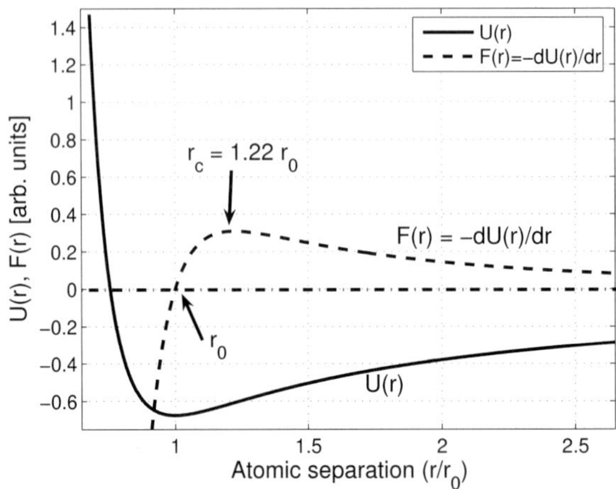

Figure B.2. *The Lennard-Jones potential $U(r)$ and the interatomic force $F(r)$ as a function of r/r_0 for $n = 1$, $m = 9$ (positive force is attraction). The average separation r_0 is found for $F = 0$, the critical separation r_c is found for $dF/dr = 0$. Both are indicated in the graph.*

atoms together becomes smaller with increasing separation: the atomic bond ruptures if the interatomic separation becomes larger than r_c. The critical separation is found by calculating the condition $dF/dr = 0$. Doing so, r_c is given by:

$$\frac{r_c}{r_0} = \left(\frac{m+1}{n+1}\right)^{\frac{1}{m-n}} \tag{B.4}$$

Filling in $n = 1$, $m = 9$ (representative of ceramic materials) results in $r_c/r_0 = 1.22$. Thus, the specimen will break if the strain $\epsilon = \Delta l/l > 22\%$. In a homogeneous, isotropic material, the relation between stress and strain is given by Young's formula $\sigma = \epsilon E$, with E Young's modulus. Thus, there is a critical stress σ_c corresponding to the critical atomic separation r_c, and if σ_c is exceeded, the material will break. In practice ceramics fracture at much lower strain than the 22% estimated from (B.4). In fact, in practice ceramics hardly deform before they break. The discrepancy is caused by the behavior of small flaws in the crystals making up the ceramics. Suppose a ceramic specimen subject to a uniform stress σ_{yy} in the y-direction. In a homogeneous and isotropic material, a uniform stress causes a uniform strain $\epsilon_{yy} = \sigma_{yy}/E$. Suppose the specimen has an elliptical flaw in it with size $2c$ in x-direction and $2b$ in y-direction. The stress field at the tips of the ellipse is complex, but a simple relation exists for the stress at the tip of the ellipse [Cook, 1994]:

$$\sigma_{yy,\text{tip}} = \sigma\left(\frac{2c}{b}\right) \tag{B.5}$$

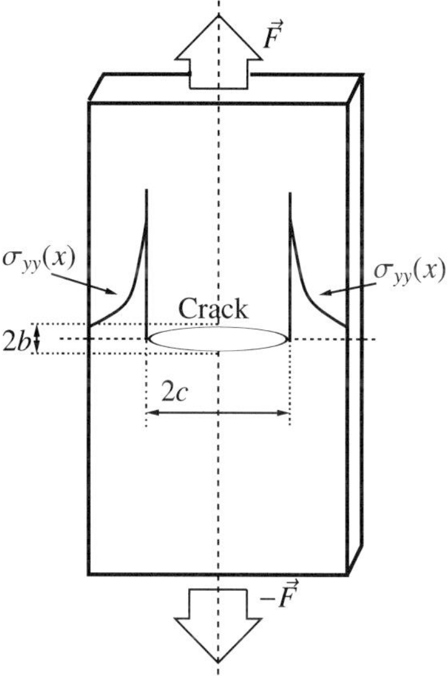

Figure B.3. *Ceramic body subject to uniaxial load $\vec{F}$, causing a stress $\sigma_{yy}(x)$ in the body. At the tips of the ellipse the magnitude of $\sigma_{yy}(x)$ increases rapidly. If σ_c is exceeded at the tip of the crack, the atomic layer will come apart, extending the crack, until eventually the specimen fails completely.*

For a slender ellipse ($c \gg b$) the stress at the tip can easily exceed σ_c even if the applied stress σ_{yy} is on average smaller than σ_c. The stress required to start propagation of a crack of arbitrary form is called the fracture stress σ_f. It is the maximum uniform stress a cracked specimen can endure without breaking. The derivation of σ_f will not be given here, just the result [Cook, 1994, Wachtman, 1996]:

$$\sigma_f = \sqrt{\left(\frac{2E\gamma}{c\psi}\right)} \tag{B.6}$$

with γ the surface energy per unit area of the material, c a characteristic dimension of the crack, and ψ a factor depending on the shape of the crack. From the previous equations, some important conclusions can be drawn:

- If the σ_c is exceeded for any crack in the material, the crack will start to propagate, increasing c. This decreases the critical stress required to separate the atomic layers ((B.6): the crack will grow without bound and the specimen fails.
- The strength of fabricated ceramics cannot be known exactly a-priori, because the strength of the specimen depends on the ab-initio crack sizes and distribution. This also implies that (mechanical) testing of ceramic products is difficult.

The fracture mechanism described leads to instantaneous failure of the specimen if the fracture stress σ_f is exceeded. This phenomenon is called *unstable crack propagation*. A related phenomenon is *stable crack propagation*. Assume we have a ceramic body with a space-dependent, non-uniform stress field within it. If a crack starts to propagate, it can propagate into a region where the stress no longer exceeds σ_c, and the crack will not propagate further. Examples of space dependent stress fields are stress fields around inclusions, stress fields where two specimens are joined together, and stress field caused by heating up or cooling down.

B.3 Thermal behavior

In the next section, the effects of rapid temperature changes on ceramic strength will be introduced. It will be shown that the strength of ceramics can be permanently decreased by thermal shock.
Ceramic materials are produced at an elevated temperature, and cooling down causes space dependent stress fields in the material. As a result, the produced ceramic will contain stable cracks, i.e. cracks which have propagated up to the point where they no longer grow. Temperature dependent stress fields originate from thermal expansion of the material. This is formalized as:

$$\frac{\Delta L}{L} = \alpha(T - T_0) \tag{B.7}$$

with α the thermal expansion coefficient. If a body is constrained, i.e. the body cannot move, a temperature change from T_0 to T will induce an effective strain in the body $\epsilon = \alpha(T - T_0)$. This will cause a stress in the body:

$$\sigma = \epsilon E = E\alpha(T - T_0) \tag{B.8}$$

In a body subject to a temperature change, the magnitude of the temperature change is generally not uniform throughout the material, inducing a space dependent stress field. If the induced stress is large enough, the stable cracks will start to propagate. The temperature difference required to start crack propagation can be calculated. The derivation will not be given here, just the result for a fully constrained body, cooled uniformly with a temperature change $T - T_0 = \Delta T$, with in it a distribution of N cracks per unit volume with a characteristic dimension c. The critical temperature change to start crack propagation ΔT_c is given by [Wachtman, 1996]:

$$\Delta T_c = \sqrt{\frac{\pi\gamma(1-2\nu)^2}{2E_0\alpha^2(1-\nu^2)}}\left(1 + \frac{16\left(1-\nu^2\right)Nc^3}{9(1-2\nu)}\right)\sqrt{\frac{1}{c}} \tag{B.9}$$

with ν Poisson's ratio. If ΔT_c is exceeded in a ceramic body, e.g. if the outside cools rapidly while the center remains hot, the cracks in the material will start to propagate. If the temperature difference is just above ΔT_c, the cracks will propagate just a bit, and the material will be

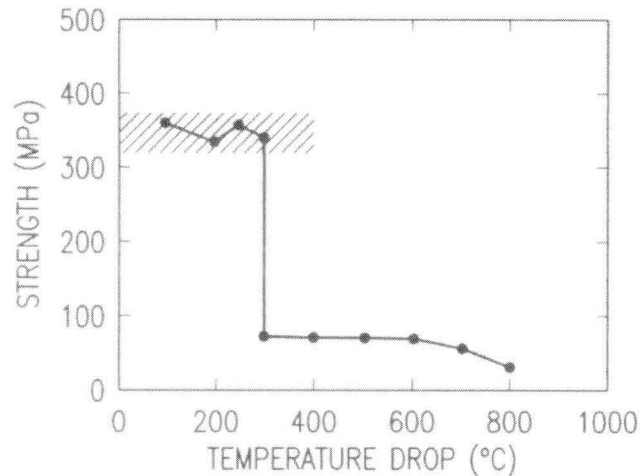

Figure B.4. *Illustration of the effect of thermal shock. If the ceramic body is cycled through a thermal cycle with ΔT_c not too large (the shaded area), the strength of the specimen remains the same after the thermal cycle is repeated. However, if the specimen is exposed to a thermal cycle larger than ΔT_c only once, the strength is permanently reduced. In this case, ΔT_c is close to 300 °C.*

left with slightly larger cracks than before. Larger cracks weaken the material, and this effect is known as thermal shock: a ceramic can be cycled through many thermal cycles with unchanged strength if $\Delta T < \Delta T_c$, but if ΔT_c is exceeded only once, the material is permanently weakened. Thermal shock is illustrated in figure B.4. If a ceramic is subject to a $\Delta T \gg \Delta T_c$, the cracks will propagate without bound and the specimen will fail.

B.4 Plasticity

All materials deform when a stress is induced. If the stress is small, the material will regain its original shape when the stress is taken away. This is known as elastic deformation. If the stress exceeds a certain threshold, the material will not regain its original shape after the stress is taken away and the specimen is permanently deformed: plastic deformation. If an even larger stress is induced the material will ultimately fail. Elastic and plastic deformation are commonly encountered in metals. Ceramics also show elastic deformation, but they fail long before the plastic deformation regime is entered. This behavior can be explained from the atomic lattice making up a ceramic. In figure B.5 a drawing is given of atoms in a lattice. In ceramics the atomic bonds are covalent and/or ionic, meaning that the atoms in the lattice exchange one or more electrons. This causes the atoms to have an effective electric charge. Plasticity is ultimately caused by slip of planes of atoms. In figure B.5 the opposite electric charges cause slip in horizontal and vertical directions to become very difficult. There is only one slip direction, namely under an angle of 45°. This is unlike metal lattices, where the atoms all have the same charge with the electrons moving in between them, making slip easier. Dislocations also play in important role, both in metals and ceramics. In fact, in metals most slip is enabled by the presence of dislocations. In ceramics dislocations are also present, but their movement is obstructed by the fact that slip directions are limited.

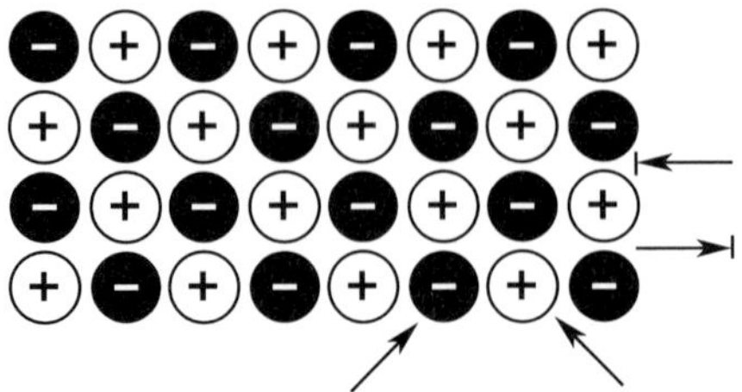

Figure B.5. *A simple atomic lattice, with '+' and '-' indicating positive and negative electric charge. Horizontal movement of the layers is difficult because ions with identical charge come closer together. Movement is possible under an angle of 45°, because then the atomic layers can slip more easily.*

As theorized above, plasticity may be possible in a single crystal in some directions, and this is observed in practice, for instance in Al_2O_3 single crystals. Most ceramics are composed of many randomly ordered crystals (polycrystalline), so plastic deformation by slip of atomic planes and/or dislocations, is almost impossible. At elevated temperatures the thermal motion of the atoms in the lattice enables some slip, and this is known as the brittle to ductile transition. For most engineering ceramics, the brittle to ductile transition occurs at high temperatures, generally $> 0.5T_{\mathrm{melt}}$, and even then plasticity is limited. The brittle to ductile transition also occurs in metals, e.g. in cast iron, which is brittle at low temperatures.

B.5 Fiber reinforced ceramics

As stated earlier, fracture of ceramics is governed by the cracks within the material. To improve the strength of ceramics, they can be reinforced by fibers, whiskers or platelets. The fibers need not be of the same composition as the ceramic, e.g. SiC fibers in Al_2O_3. The fibers, whiskers and platelets improve the mechanical properties by (see figure B.6):

- Crack deflection: if a propagating crack finds a whisker in its path, the crack cannot propagate further easily.
- Crack bridging: even if a crack exists in the material, the whisker can still tie the two parts together.
- Whisker pull-out: if a bridged crack becomes wider, the whisker will eventually be pulled out of the surrounding material, or simply break. In both cases energy is dissipated.

For this reason, fiber- and whisker reinforced ceramics are the structural material of choice for the Generation IV GCFR. For example, a fiber-reinforced SiC structure can be made in the following way: using SiC (or carbon) fibers a 3-D structure is woven. This 3-D object is then infiltrated with a fluid or gas that will produce SiC in a chemical reaction. The result is a very strong, 3-D SiC fiber-reinforced product.

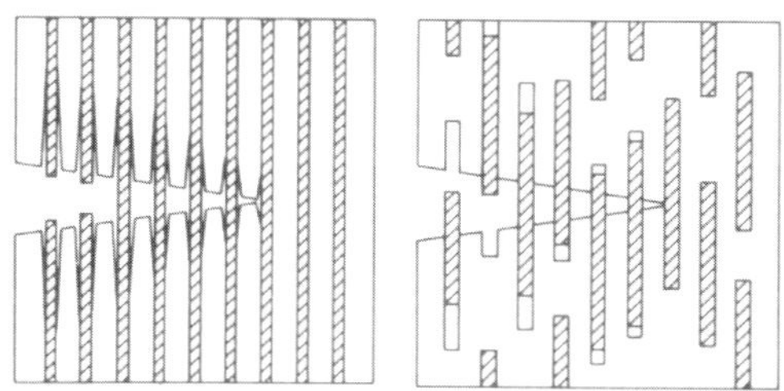

Figure B.6. *Fiber reinforced ceramics. If the crack propagates, the fibers have to be broken or pulled out of the surrounding ceramic. This makes crack propagation more difficult, increasing the strength of the material. In the left figure a layered structure is made with long fibers (shaded). In the right figure small whiskers are dispersed throughout the material to improve strength.*

B.6 Interaction of neutrons with ceramics

Irradiation with neutrons causes material damage. The most important effects are caused by collisions between (possibly highly energetic) neutrons and atoms in the crystal lattice. In a high energy collision, enough energy can be transferred from the neutron to the atom to push the atom out of its lattice site. The moving atom can cause more damage in the crystal lattice by collisions with other atoms. The final result is a vacancy in the lattice, with the atom usually ending up as an interstitial. As a result, the material properties will change, and volume changes might occur. These are well known and important effects in all nuclear reactors, especially in fast reactors with their hard neutron spectrum.

Irradiation experience with ceramic materials is mostly limited to UO_2 fuel. Structural materials for reactors are usually metallic or graphite. Although graphite can be considered a ceramic, modeling of irradiation damage is still largely empirical and highly material specific. For application of ceramics on a large scale as structural materials in a fast reactor, much development will have to be done.

Bibliography

J.P. Ackermann, T.R. Johnson, L.S.H. Chow, E.L. Carls, W.H. Hannum, and J.J. Laidler. Treatment of wastes in the IFR fuel cycle. *Progress in Nuclear Energy*, 31(1/2):141–154, 1997.

ANSI/ANS-5.1-1994. W2004: Decay heat power in light water reactors. Technical report, ANSI/ANS, 2004.

H. Bailly, D. Ménessier, and C. Prunier. *The nuclear fuel of pressurized water reactors and fast reactors*. Lavoisier Publishing Inc., 1999.

B.L. Broadhead, B.T. Rearden, C.M. Hopper, J.J. Wagschal, and C.V. Parks. Sensitivity- and uncertainty-based criticality safety validation techniques. *Nuclear Science and Engineering*, 146:340–366, 2004.

P.N. Brown, A.C. Hindmarsh, and G.D. Byrne. *DVODE: Variable-coefficient ordinary differential equation solver with fixed-leading-coefficient implementation*, 1989. Available online at http://www.netlib.org/ode/.

D.G. Cacuci. *Sensitivity and uncertainty analysis*, volume I, Theory. CRC Press, 2003.

J. Chermanne. GCFR fuel assemblies. In *Gas-cooled fast reactors*, IAEA-TECDOC-154. IAEA, Vienna, Austria, 1972a.

J. Chermanne. Past and future programmes of the GBR Association. In *Gas-cooled fast reactors*, IAEA-TECDOC-154. IAEA, Vienna, Austria, 1972b.

J. Chermanne and P. Burgsmüller. Gas-cooled breeder reactor safety. In *Gas-cooled reactor safety and licensing aspects*, IWGGCR-1. IAEA, Vienna, Austria, 1981.

H. Choi and C.J. Park. A physics study on thorium fuel recycling in a CANDU reactor using dry process technology. *Nuclear Technology*, 153:132–145, 2006.

M. Cometto, P. Wydler, and R. Chawla. A comparative physics study of alternative long-term strategies for closure of the nuclear fuel cycle. *Annals of Nuclear Energy*, 31:413–429, 2004.

A. Conti and J.C. Bosq. 01/2004 600 MWth GFR cores containing plates CERCER characteristics. Technical report, CEA, 2004.

R.F. Cook. *Structure and properties of ceramics*, volume 11 of *Materials science and technology series*, chapter 7. VCH Verlagsgesellschaft mbH, Weinheim, Germany, 1994.

M. Dalle Donne and C.A. Goetzmann. Gas cooled fast breeder reactor design, development, and safety features. In *Gas-cooled reactors: HTGR and GCFBR*. ANS, Hinsdale, USA, 1974.

W.J.M. de Kruijff and A.J. Janssen. On the definition of the fuel temperature coefficient of reactivity for pin-cell calculations on an infinite lattice. *Annals of Nuclear Energy*, 20: 639–648, 1993.

DIN. Berechnung der Nachzerfallsleistung der Kernbrennstoffe von Leichtwasserreaktoren; Nichtrezyklierte Kernbrennstoffe. Technical Report DIN 25463-1, Deutsches Institut für Normung e.V., May 1990.

J.J. Duderstadt and L.J. Hamilton. *Nuclear Reactor Analysis*. John Wiley & Sons, Inc, 1976.

P. Dumaz, A. Conti, C. Bassi, and C. Poette. Personal communications. CEA Cadarache, 2006.

D.P. Every. Silicon carbide clad fuel pins - UK manufacturing & irradiation experience, June 2006. GCFR-STREP Deliverable D19. Draft version, final issue available in 2009.

J.C. Garnier, N. Chauvin, P. Anzieu, G. Francois, T.Y.C. Wei, T. Taiwo, M. Meyer, P. Hejzlar, H. Ludewig, and A. Baxter. Feasibility study of an advanced GFR design - Trends and safety options - status of France & US studies. In *GLOBAL 2003*, New Orleans, USA, November 16-20 2003. ANS.

I.C. Gauld, O.W. Hermann, and R.M. Westfall. *ORIGEN-S: SCALE system module to calculate fuel depletion, actinide transmutation, fission product buildup and decay, and associated radiation source terms*. Oak Ridge National Laboratory, April 2005.

R. Gontard and H. Nabielek. Performance evaluation of modern htr triso fuels. Technical Report HTA-IB-05/90, Forschungszentrum Jülich GmbH, 1990.

E. Guyon, J.-P. Hulin, L. Petit, and C.D. Mitescu. *Physical Hydrodynamics*. Oxford University Press, 2001.

C.S. Handwerk, M.J. Driscoll, and P. Hejzlar. Use of beryllium oxide to shape power and reduce void reactivity in gas cooled fast reactors. In *PHYSOR 2006*. ANS, September 10-14 2006.

K. Hesketh and E. Sartori. OECD/NEA activities on MOx issues. In *Physics and fuel performance of reactor-based plutonium disposition*, pages 121–136. OECD/NEA, 1999.

E.A. Hoffman and W.M. Stacey. Comparative fuel cycle analysis of critical and sub-critical fast reactor transmutation systems. *Nuclear Technology*, 144:83–106, 2003.

J.E. Hoogenboom and J.L. Kloosterman. Generation and validation of ORIGEN-S libraries for depletion and transmutation calculations based on JEF2.2 and EAF3 basic data. *Nuclear Engineering and Design*, 170:107–118, 1997.

IAEA-TECDOC-1349. Potential of thorium based fuel cycles to constrain plutonium and reduce long lived waste toxicity. Technical report, IAEA, 2003. Table on page 55.

IAEA-TECDOC-626. Safety related terms for advanced nuclear plants. Technical report, IAEA, 1991.

T. Ikegami. Core concept of compound process fuel cycle. *Progress in Nuclear Energy*, 47 (1-4):231–238, 2005.

M. Inoue, K. Ono, T. Fujioka, K. Sato, and T. Asaga. Feasibility study on Nitrogen-15 enrichment and recycling system for innovative FR cycle system with nitride fuel. In *Proceedigns of ICONE10*. ASME, USA, 2002.

M. Kambe and M. Uotani. Design and development of fast breeder reactor passive reactivity control systems: LEM and LIM. *Nuclear Technology*, 122:179–195, 1997.

Y. Kato, T. Nitawaki, and Y. Muto. Medium temperature carbon dioxide gas turbine reactor. *Nuclear Engineering and Design*, 230:195 – 207, 2004.

W.B. Kemmish, M.V. Quick, and I.L. Hirst. The safety of CO_2 cooled breeder reactors based on existing gas cooled reactor technology. *Progress in Nuclear Energy*, 10:1–17, 1981.

D. Kinii and O. Levenspiel. Fluidized reactor models. 1. For bubbling beds of fine, intermediate and large particles. 2. For the lean phase: freeboard and fast fluidization. *Ind. Chem. Eng. Res.*, 29:1226–1234, 1991.

J.L. Kloosterman. Integral validation and decay heat standards. Lectures notes FJ/OH Summer School, 2000.

J.L. Kloosterman and E.E. Bende. Plutonium recycling in pressurized water reactors: influence of the moderator-to-fuel ratio. *Nuclear Technology*, 130:227–241, 1999.

J.L. Kloosterman and J.C. Kuijper. VAREX, a code for variational analysis of reactivity effects: description and examples. In *PHYSOR 2000*, Seoul, South Korea, October 7-10 2000. ANS.

M. Konomura, T. Saisuga, T. Mizuno, and Y. Ohkubo. A promising gas-cooled fast reactor concept and its R&D plan. In *GLOBAL 2003*, New Orleans, USA, November 16-20 2003. ANS.

A. Koster, H.D. Matzner, and D.R. Nicholsi. PBMR design for the future. *Nuclear Engineering and Design*, 222:231–245, 2003.

K. Kugeler and R. Schulten. *Hochtemperaturreaktortechnik*. Springer-Verlag, 1989.

K. Kunitomi, S. Katanishi, S. Takada, T. Takizuka, and X. Yan. Japan's future - the GTHTR300. *Nuclear Engineering and design*, 233:309–327, 2004.

G. Lavialle. *CATHARE 2 v2.5 mod3.1 User's manual*. CEA, 2006. SSTH/LDAS/EM/2005-035.

J. Leppanen. Preliminary calculations on actinide management using advanced PWR MOx technology. Technical report, VTT, 2005.

Günter Lohnert. Topical issue on China's HTR-10. *Nuclear Engineering and Design*, 218 (1-3), 2002.

Günter Lohnert. Topical issue on Japan's HTTR. *Nuclear Engineering and Design*, 233 (1-3), 2004.

J.T. Long. *Engineering for nuclear fuel reprocessing*. ANS, 1978.

P.E. MacDonald. NGNP preliminary point design - results of the initial neutronics and thermal-hydraulic assessments. Technical report, Idaho National Engineering and Environmental Laboratory, 2003.

D.G. Martin. Considerations pertaining to the achievement of high burn-ups in HTR fuel. *Nuclear Engineering and Design*, 213:241–258, 2002a.

D.G. Martin. Personal communication. HTR-ECS, 2002b.

L. Massimo. *Physics of high temperature reactors*. Pergamon Press, 1976.

G. Melese and R. Katz. *Thermal and Flow Design of Helium-Cooled Reactors*. ANS, 1984.

H. Mochizuki, T. Izaki, K. Takitani, H. Koike, Y. Kobayashi, Y. Matsuki, T. Ooka, R. Tanaka, T. Watanabe, and T. Kobayashi. Design study of a He gas-cooled fast breeder reactor. In *Gas-Cooled Fast Reactors*, IAEA-TECDOC-154. IAEA, Vienna, Austria, 1972.

F. Morin, T. Cadiout, C. Poette, A. Tosello, and Y. Enuma. Preliminary design of ETDR D10 Task 1.2. Technical report, CEA, 2006. GCFR-STREP Deliverable D10.

V.B. Nesterenko, V.F. Zelensky, L.I. Kolykhan, G.V. Karpenko, V.S. Krasnorutsky, V.P. Isakov, V.P. Ashikhmin, and L.N. Permyakov. Problems of creating fuel elements for fast gas-cooled reactors working on N_2O_4-dissociating coolant. In *Gas-cooled reactor fuel development and spent fuel treatment*, IWGGCR-8. IAEA, Vienna, Austria, 1983.

OECD Nuclear Energy Agency. PBMR coupled neutronics/thermal hydraulics transient benchmark - The PBMR-400 core design - Benchmark description, September 2005. Draft-V03.

SCALE: a modular code system for performing standardized computer analyses for licensing evaluations, NUREG/CR-0200, version 4.4a. Oak Ridge National Laboratory, March 2000.

SCALE: A modular code system for performing standardized computer analyses for licensing evaluations, ORNL/TM-2005/39, version 5, Vols I-III. Oak Ridge National Laboratory, April 2005. Available from Radiation Safety Information Computational Center at Oak Ridge National Laboratory as CCC-725.

K.O. Ott and R.J. Neuhold. *Introductory Nuclear Reactor Dynamics.* ANS, 1985.

S. Permana, N. Takaki, and H. Sekimoto. Impact of different moderator ratios with light and heavy water cooled reactors in equilibrium states. *Annals of Nuclear Energy*, 33:561–572, 2006.

H.O. Pierson. *Handbook of refractory carbides and nitrides.* Noyes publications, 1996.

P. Pohl. The importance of the AVR Pebble-Bed Reactor for the future of nuclear power. In *PHYSOR 2006*, Vancouver, Canada, September 10 - 14 2006. ANS.

G. Rimpault, D. Plisson, J. Tommasi, R. Jacqmin, J.-M. Rieunier, D. Verrier, and D. Biron. The ERANOS code and data system for fast reactor neutronic analyses. In *PHYSOR 2002*, Seoul, South Korea, October 7-10 2002. ANS.

R. Roest. *Inleiding mechanica.* Delftse Universitaire Pers, 1987.

J. Rouault, A. Judd, J.C. Lefèvre, and G. Mühling. CAPRA preliminary feasibility studies, synthesis and conclusions. Technical report, CEA, 1994.

H.J. Rütten and J.C. Kuijper. Plutonium (1st and 2nd generation) cell burnup benchmark specification. Technical Report HTR-N1-02/06-s-3.1.1, European Commission, 2003.

K. Ryu and H. Sekimoto. A possibility of highly efficient uranium utilization with a pebble bed fast reactor. *Annals of Nuclear Energy*, 27:1139–1145, 2000.

M. Salvatores. Nuclear fuel cycle strategies including partitioning and transmutation. *Nuclear Engineering and Design*, 235:805–816, 2005.

M. Salvatores. *Handbook of Nuclear Reactor Calculations*, volume III, chapter "Fast Reactor Calculations", pages 263 – 363. CRC Press, 1986.

R.H. Simon, J.B. Dee, and W.I. Morita. Gas-cooled fast breeder reactor demonstration plant. In *Gas-cooled reactors: HTGR and GCFBR*, pages 336–354, Gatlinburg, USA, May 7-10 1974. ANS.

W.M. Stacey. *Nuclear Reactor Physics.* John Wiley & Sons, 2001.

R.E. Sunderland, E.K. Whyman, H.M. Beaumont, and D.P. Every. A gas-cooled dedicated minor actinide burning fast reactor: initial core design studies. In *GLOBAL 1999*, Jackson Hole, USA, August 29 - September 3 1999. ANS.

T.A. Taiwo, T.Y.C. Wei, E.E. Feldman, E.A. Hoffman, M. Fatone, J.W. Holland, I.G. Prokofiev, W.S. Yang, G. Palmiotti, R.N. Hill, M. Todosow, M. Salvatores, and A. Gandini. Particle-bed Gas Cooled Fast Reactor (PB-GCFR) Design, Final Technical Report, Sept 2001 - Aug 2003. Technical report, Argonne National Laboratory, 2003.

T.A. Taiwo, E.A. Hoffman, R.N. Hill, and W.S. Yang. Evaluation of long-life transuranics breakeven and burner cores for waste minimization in a PB-GCFR fuel cycle. *Nuclear Technology*, 155(1):55 – 66, 2006.

A. Talamo and W. Gudowski. Performance of the gas turbine-modular helium reactor fueled with different types of fertile TRISO particles. *Annals of Nuclear Energy*, 32(16):1719–1749, 2005.

C. Tang, Y. Tang, J. Zhu, Y. Zou, J. Li, and X. Ni. Design and manufacture of the fuel element for the 10 MW High Temperature Gas-cooled Reactor. *Nuclear Engineering and Design*, 218:91–102, 2002.

K. Tasaka, J. Katakura, H. Ihara, T. Yoshida, S. Iijima, R. Nakasima, T. Nakagawa, and H. Takano. JNDC Nuclear data library of fission products, second version. Technical Report JAERI-1320, JAERI, 1990.

A. Torri and D.R. Buttemer. Gas-cooled fast reactor safety - an overview and status of the U.S. program. In *Gas-cooled reactor safety and licensing aspects*, IWGGCR-1. IAEA, Vienna, Austria, 1981.

U.S. DOE Nuclear Energy Research Advisory Committee and the Generation IV International Forum. A technology roadmap for Generation IV nuclear energy systems, December 2002. Available online at http://gif.inel.gov/roadmap/.

W.F.G. Van Rooijen, J.L. Kloosterman, T.H.J.J. van der Hagen, and H. van Dam. Fuel design and core layout for a gas cooled fast reactor. *Nuclear Technology*, 151:221–238, 2005.

K. Verfondern, J. Sumita, S. Ueta, and K. Sawa. Modeling of fuel performance and metallic fission product release behavior during HTTR normal operating conditions. *Nuclear Engineering and Design*, 210:225–238, 2001.

A.H.M. Verkooijen and J.W. de Vries. Experience with MTR fuel at the HOR reactor. *Kerntechnik*, 69:88–91, 2004.

J.B. Wachtman. *Mechanical properties of ceramics*. John Wiley & Sons, 1996.

A.E. Waltar and A.B. Reynolds. *Fast Breeder Reactors*. Pergamon Press, 1981.

L.C. Walters. Thrity years of fuels and materials information from EBR-II. *Journal of Nuclear Materials*, 270:39–48, 1999.

M.L. Williams. *Handbook of Nuclear Reactor Calculations*, volume III, chapter "Perturbation theory for nuclear reactor analysis", pages 63 – 188. CRC Press, 1986.

List of Symbols

Throughout this thesis scalar quantities and operators are indicated by a simple symbol, vector quantities are indicated by an arrow sign (e.g. Σ_a vs $\vec{\Sigma}_a$). Matrix quantities are indicated by an underline, e.g. $\underline{M}$. Adjoint quantities and operators are indicated by '*'. Brackets '$\langle , \rangle$' indicate inner products, i.e. integration over all continuous variables, summation over all integer variables.

Abbreviations

AIROX Atomics International Reduction Oxidation

BOC Beginning of Cycle

CAPRA Concept to Amplify Plutonium Reduction in Advanced fast reactors

CDS Constant Decay Source

CEA Commisariat à l'Energie Atomique

CR Control Rod

CSD Control / Shut Down rod

DHR Decay Heat Removal

DSD Diverse Shut Down rod

EOC End of Cycle

EPR European Pressurized water Reactor (substitute 'Evolutionary' if on North American continent)

EU FP6 European Union 6th Framework Program

eV Electronvolt

FIMA Fissions per Initial Metal Atom

FTC Fuel Temperature Coefficient

GCFR Gas Cooled Fast Reactor

gHM gram Heavy Metal

HM Heavy Metal, i.e. uranium and all heavier elements

HTR High Temperature Reactor

LIM Lithium Injection Module

LMFBR Liquid Metal Fast Breeder Reactor

LOCA Loss Of Coolant Accident

LOFA Loss Of Flow Accident
LWR Light Water Reactor
MA Minor Actinides
MWe Megawatt electric
MWth Megawatt thermal
PK Point Kinetics
PUREX Plutonium Uranium REduction and eXtraction
PWR Pressurized Water Reactor
RCP Random Close Packing
RG Reactor Grade (plutonium)
RPV Reactor Pressure Vessel
SNF Spent Nuclear Fuel
STREP Specific Targeted REsearch Program
WG Weapons Grade (plutonium)

Nuclear quantities

α Capture to fission ratio
α_i Data element appearing in operator
β^- Radioactive beta-decay
$\beta_i(t), \beta(t)$ Delayed neutron yield of group i, total effective delayed neutron yield
$\chi(E)$ Fission spectrum
$\Delta\cdot$ Perturbation on $\cdot$
$\delta(t - t_0)$ Dirac delta function
$\epsilon_{jk}(t)$ Pseudo decay heat function
η Average number of new neutrons per absorption
γ_{jk} Decay heat contribution of isotope j, in group k
$\hat{\Omega}$ Spatial angle / direction (unit vector)
λ_0, λ Eigenvalue of Boltzmann equation
$\Lambda_0, \Lambda(t)$ Steady state, transient neutron generation time
λ_i Delayed neutron precursor decay constants
λ_x Decay constant for radioactive decay mode x
μ_{jk} Decay constant for decay heat, of isotope j, in group k
ν Average number of new neutrons per fission
ϕ, ϕ_0 Neutron flux / unperturbed flux
$\psi(t)$ Time dependent fission rate
ρ_0, ρ Reactivity

σ_a, Σ_a	Absorption cross section (microscopic, macroscopic)
σ_c, Σ_c	Capture cross section (microscopic, macroscopic)
σ_f, Σ_f	Fission cross section (microscopic, macroscopic)
σ_s, Σ_s	Scatter cross section (microscopic, macroscopic)
σ_t, Σ_t	Total cross section (microscopic, macroscopic)
σ_x	Cross section for reaction type x (microscopic)
$\zeta_i(t)$	Delayed neutron precursor concentrations
$\vec{b}$	Response selection vector
B_g	Geometric buckling
BG	Breeding Gain
BR	Breeding Ratio
C_c	Constant for conservative decay heat calculations
E	Energy
E_{rel}	Average energy release per fission
$F_0, F(t)$	Steady state, transient fission rate
FD	Fissile mass destroyed per cycle
FG	Fissile mass gained per cycle
f_j	Fraction of fissions in isotope j for decay heat calculation
FP	Fissile mass produced per cycle
$f(t)$	Decay heat impulse response
h	Fuel performance parameter
k_{eff}	Effective multiplication factor
k_∞	Infinite medium multiplication factor
L_0, L	Neutron loss operator
L_D	Neutron diffusion length
$\underline{M}$	Transmutation matrix
m	Mass
N	Nuclide density
n	Neutron
n_0	Number of HM nuclides in kernel at BOC
$\vec{N}_{\text{cool}}$	Fuel composition after cool down
$\vec{N}_{\text{feed}}$	Composition of feed material for new fuel
N_{feed}	Amount of feed material to make new fuel
$\vec{N}_{\text{ini}}$	Initial fuel composition
$\vec{N}_{\text{irrad}}$	Fuel composition after irradiation

$\vec{N}_{\text{new}}$	New fuel composition, made from reprocessed material
$\vec{N}_{\text{repro}}$	Composition of reprocessed material
N_{target}	Target amount of new fuel
p	Proton
P_0, P	Neutron production operator
$P_d(t)$	Decay power
$P_n(t)$	Neutronic power
Q	Independent source (nuclides or neutrons)
$\vec{r}, r$	Space location
R	Formal response or performance parameter
R_{feed}	Performance of feed material
R_{repro}	Performance of reprocessed material
$\underline{S}$	Reprocessing efficiency
$S_{i,j}$	Sensitivity of response i to data α_j
v	Velocity
$\vec{v}_{\text{feed}}$	Composition of feed material
w	Reactivity weight of an isotope
$y_{j\rightarrow i}$	Yield of j due to fission of i
z	Number of nuclides released into buffer per fissioned metal atom
Thermalhydraulic quantities	
Δp_{pb}	Pressure drop over packed bed
ϵ	Porosity of packed bed
$\lambda_{I/II}$	Thermal conductivity of pebble fuel zone / graphite shell
μ_f	Viscosity of fluid
ρ_f	Density of fluid
τ	Circulator time constant for rundown
A	Geometrical surface of packed bed
c_p	Heat capacity
d_p	Diameter of particles making up packed bed
h	Heat transfer coefficient
h_c	Core height
h_{pb}	Height of packed bed
k	Boltzmann's constant
$\dot{m}$	Coolant mass flow rate
P	Volumetric power density (core averaged)

p_{buf} Pressure in buffer layer
q Power density pebble fuel zone
q_p Power of fuel pebble
r_{fz} Radius of pebble fuel zone
r_{in} Inside radius LIM tube
r_k TRISO kernel radius
r_{out} Outside radius LIM tube
r_{peb} Radius of pebble
r_t TRISO particle radius
$T_{I/II}$ Temperature in fuel zone / graphite shell of pebble
T_0 Steady state nominal fuel temperature
T_{buf} Average temperature buffer layer
$T_{\text{core,in/out}}$ Coolant temperature at core inlet / outlet
T_G Bulk coolant gas temperature
T_L Film temperature at pebble surface
T_m Maximum (centerline) temperature in a pebble
t_s Time since start of transient
u Superficial fluid velocity of packed bed
V_{buf} Volume of buffer layer

Ceramic properties

α Thermal expansion coefficient
ΔT_c Critical temperature difference for crack propagation
ϵ Strain
γ Surface energy per unit area
ν Poisson ratio
ψ Crack shape factor
σ Stress
σ_c Critical stress
σ_f Fracture stress
b, c Crack critical dimensions
E Modulus of elasticity
$F(r)$ Interatomic force
L Length of specimen
r Atomic separation
r_c Critical atomic separation
$U(r)$ Interatomic potential

List of publications

- W.F.G. van Rooijen and D. Lathouwers, *'Sensitivity analysis of the kinetic behavior of a Gas Cooled Fast Reactor to variations of delayed neutron parameters'*, submitted for M&C 2007 (2006)
- W.F.G. van Rooijen, J.L. Kloosterman, T.H.J.J. van der Hagen and H. van Dam, *'Passive reactivity control for a Gas Cooled Fast Reactor'*, submitted to Nuclear Technology (2006)
- W.F.G. van Rooijen, J.L. Kloosterman, T.H.J.J. van der Hagen and H. van Dam, *'Definition of breeding gain for the closed fuel cycle and application to a Gas Cooled Fast Reactor'*, accepted for publication in Nuclear Science and Engineering (2006)
- W.F.G. van Rooijen, J.L. Kloosterman, T.H.J.J. van der Hagen and H. van Dam, *'Definition of breeding gain for the closed fuel cycle and application to a Gas Cooled Fast Reactor'*, PHYSOR-2006, Vancouver, Canada
- W.F.G. van Rooijen, J.L. Kloosterman, T.H.J.J. van der Hagen and H. van Dam, *'Passive shutdown device for Gas Cooled Fast Reactor: Lithium Injection Module'*, PHYSOR-2006, Vancouver, Canada.
- A.M. Ougouag, J.L. Kloosterman, W.F.G. van Rooijen, H.D. Gougar, W.K. Terry, *Investigations of bounds on particle packing in Pebble-Bed High Temperature Reactors*, Nuclear Engineering and Design, **236**, 669 - 676 (2006)
- W.F.G. van Rooijen, J.L. Kloosterman, T.H.J.J. van der Hagen and H. van Dam, *'Fuel design and core layout for a Gas Cooled Fast Reactor'*, Nuclear Technology **151-3**, September 2005.
- J.-Y. Malo, C. Bassi, A. Conti, P. Dumaz, J.-C. Garnier, F. Morin, W.F.G. van Rooijen, *'The thermalhydraulic design of the experimental and technology demonstration reactor (ETDR)'*, NURETH-11 (paper 434), France 2005.
- W.F.G. van Rooijen, *'Modelisation and preliminary transient analysis of the ETDR start-up core using CATHARE'*, Mémo CEA, France (2004)
- W.F.G. van Rooijen, *'Introductory study of heat exchanger design and application to a conceptual design of a primary heat exchanger for ETDR'*, Note Technique CEA, France (2004)
- W.F.G. van Rooijen, J.L. Kloosterman, T.H.J.J.van der Hagen, *'A prospective HTR study'*, IRI report IRI-131-2004-003

- W.F.G. van Rooijen, J.L. Kloosterman, T.H.J.J.van der Hagen and H. van Dam, *'Fuel design and core layout for a Gas Cooled Fast Reactor'*, PHYSOR-2004, Chicago, Ill., USA (2004)
- W.F.G. van Rooijen, J.L. Kloosterman, T.H.J.J.van der Hagen and H. van Dam, *'Design of a spherical fuel element for a Gas Cooled Fast Reactor'*, Third information exchange meeting on basic studies in the field of high temperature engineering, Oarai, Japan (2003)
- K.Yamamoto, F. Asano, W.F.G. van Rooijen, E.Y. Ling, T. Yamada, N. Kitawaki, *'Estimation of the Number of Sound Sources Using Support Vector Machines And Its Application To Sound Source Separation'*, ICASSP03, Hong Kong, China (2003)
- W.F.G. van Rooijen, E.Y. Ling, F. Asano, K. Yamamoto, N. Kitawaki, *'Acoustic source number estimation using support vector machine and its application to source localization/separation system'*, Tech. rep. IEICE EA2002-41 (2002)
- W.F.G. van Rooijen, *'Distributed Mode Loudspeakers for Wave Field Synthesis'*, M.Sc. thesis, Delft University of Technology (2001)

Improving Fuel Cycle Design and Safety Characteristics of a Gas Cooled Fast Reactor

The Generation IV Forum is an international nuclear energy research initiative aimed at developing the fourth generation of nuclear reactors, envisaged to enter service halfway the 21st century. One of the Generation IV reactor systems is the Gas Cooled Fast Reactor (GCFR), the subject of study in this thesis. The Generation IV reactor concepts should improve all aspects of nuclear power generation. Within Generation IV, the GCFR concept specifically targets sustainability of nuclear power generation.

In all nuclear reactors non-fissile material is converted to fissile fuel (e.g. ^{238}U to ^{239}Pu conversion). If the neutrons inducing fission are highly energetic, the opportunity exists to convert more than one non-fissile nucleus per fission, thereby effectively breeding new nuclear fuel. Reactors operating on this principle are called 'Fast Breeder Reactor', 'fast' because of the equivalence between speed and energy. Since natural uranium contains 99.3% of the non-fissile isotope ^{238}U, breeding increases the energy harvested from the nuclear fuel. If nuclear energy is to play an important role as a source of energy in the future, fast breeder reactors are essential for breeding nuclear fuel. Fast neutrons are also more efficient to destruct heavy isotopes, such as neptunium, americium and curium isotopes (the so-called Minor Actinides, MA). Since these Minor Actinides dominate the long-term radioactivity of nuclear waste, the waste life-time can be shortened if the MA nuclei are destroyed. An important prerequisite of sustainable nuclear energy is the closed fuel cycle, where only fission products are discharged to a final repository, and all Heavy Metal (HM) are recycled. Thus the reactor should breed just enough fissile material to allow refueling of the same reactor, adding only fertile material to the recycled material. Other key design choices for GCFR are highly efficient power conversion using a direct cycle gas turbine, and better safety through the use of helium, a chemically inert coolant which cannot have phase changes in the reactor core.

Because the envisaged GCFR core temperatures and operating conditions are similar to thermal-spectrum High Temperature Reactor (HTR) concepts, the research for this thesis initially focused on a GCFR design based on existing HTR fuel technology, i.e. coated particle fuel, assembled into fuel assemblies. It was found that such a fuel concept could not meet the Generation IV criteria set for GCFR: self-breeding is difficult, the temperature gradients within the fuel assemblies would be too high, and fuel economy is poor. As a solution, two improved fuel concepts are proposed for GCFR, one being a redesign of the classic TRISO coated particle fuel, and the other one being an innovative hollow sphere design. Both fuel elements are used in a core design based on direct cooling of the coated particle fuel. To increase the neutronic margins and obtain adequate self-breeding capabilities, the proposed reactor has 2400 MWth power output and a power density of 50 MW/m^3. With both types of coated particle fuel, it is possible to obtain the closed fuel cycle. The MA loading of the fuel remains rather limited (about 1%). Long irradiation intervals (several years) are possible with a low burnup reactivity swing, which reduces the required over-reactivity of the fresh core and reduces control rod requirements during operation.

In the closed fuel cycle it is important to be able to predict whether a certain initial fuel composition will in fact yield a new fuel, after irradiation, cool down and reprocessing, with which the reactor can be restarted. A theoretical framework is presented in this thesis which allows calculation of the 'Breeding Gain' of the reactor. The *BG* quantifies the performance of the fuel for batch $i + 1$ as a function of the composition of the initial fuel of batch i. If this *BG* can be made equal to zero, both fuel compositions give the same nuclear performance. To be able to calculate the fuel performance, the reactivity weight, i.e. the contribution of each isotope to the overall reactivity of the reactor, needs to be estimated. It is proposed in this thesis to calculate these reactivity weights using a first-order *eigenvalue* perturbation calculation. It is shown that this approach yields an expression which reduces to a well-established formula for reactivity weights. All steps in the fuel cycle, i.e. irradiation, cool down and reprocessing, have to be taken into account to calculate the Breeding Gain for the closed fuel cycle. First order *nuclide* perturbation theory provides an efficient method to calculate the effects of small variations of the initial fuel composition on the performance of the closed fuel cycle. The theory is applied to the closed fuel cycle of a 600 MWth Gas Cooled Fast Reactor. The result is that the closed fuel cycle can be obtained if the reprocessing is efficient enough in retrieving the transuranics from the irradiated fuel ($> 99\%$). Calculations were done adding extra MA to the GCFR fuel, to estimate the transmutation potential of the GCFR concept. Extra Minor Actinides in the fuel improve the Breeding Gain, and reduce the burnup reactivity swing.

The Gas Cooled Fast Reactor core power density is high in comparison to other gas cooled reactor concepts. Like all nuclear reactors, the GCFR produces decay heat after shut down, which has to be transported out of the reactor under all circumstances. The layout of the primary system therefore focuses on using natural convection Decay Heat Removal (DHR) where possible, with a large coolant fraction in the core to reduce friction losses. However, due to the combination of high power density and low thermal inertia in the core, transients in the GCFR core may lead to high temperatures. To protect the reactor under all circumstances during transients, passive reactivity control devices are researched. These devices control the reactor power under off-nominal conditions when all other control devices fail. The proposed devices use liquid ^6Li as an absorber, which is passively introduced into the core. Activation of the device is by freeze seals, which melt when the core outlet temperature is too high. These devices can be integrated into the normal control assemblies of the reactor while still keeping enough room available for the regular control elements. The passive devices are shown to adequately limit the power production of the GCFR core. It is also shown that natural circulation DHR is possible under pressurized core conditions.

Delft, October 2006,

W.F.G. van Rooijen

Splijtstofcyclus en veiligheidskarakteristieken van een Snelle Gasgekoelde Reactor

Het Generation IV Forum is een internationaal onderzoeksprogramma op het gebied van kernenergie, met als doel het ontwikkelen van de vierde generatie kernreactoren, die vanaf halverwege de 21ste eeuw in bedrijf zouden moeten komen. Een van de reactortypen binnen Generation IV is de Snelle Gasgekoelde Reactor (SGR). De geselecteerde reactorconcepten voor Generation IV moeten alle aspecten van kernenergie verbeteren. Het specifieke doel van de SGR binnen Generation IV is duurzaamheid.

In elke kernreactor wordt niet-splijtbaar materiaal omgezet in splijtbaar materiaal, bijv. ^{238}U in ^{239}Pu. Als de neutronen die kernsplijting induceren, een voldoende hoge energie hebben, bestaat de mogelijkheid om meer dan één splijtbare kern te produceren voor iedere verspleten kern. Op deze wijze wordt splijtbaar materiaal gekweekt. Reactoren die werken volgens dit principe worden 'Snelle Kweekreactor' genoemd, 'snel' vanwege de equivalentie tussen energie en snelheid. Aangezien natuurlijk uranium voor 99.3% uit de niet-splijtbare isotoop ^{238}U bestaat, kan door kweken de totale energieopbrengst van de nucleaire spijtstof sterk vergroot worden. Opdat kernenergie in de toekomst een belangrijke rol kan spelen in de energievoorziening, zijn snelle reactoren essentieel, vanwege het kweken van splijtstof. Snelle neutronen zijn tevens efficiënter voor het transmuteren van zware isotopen, zoals de isotopen van neptunium, americium and curium (hier aangeduid als 'actiniden'). Aangezien de transuranen op de lange termijn de radioactiviteit van kernafval domineren, kan de levensduur van kernafval sterk verkort worden als de actiniden verspleten worden. Een belangrijke voorwaarde voor duurzame toepassing van kernenergie is de gesloten spijtstofcyclus, waarbij alleen splijtingsproducten naar een (geologische) eindberging gaan, en alle zware metalen (uranium en zwaardere elementen) gerecycleerd worden. Tijdens bedrijf moet de reactor zoveel splijtstof kweken dat het mogelijk is een nieuwe lading spijtstof te maken door alleen kweekmateriaal toe te voegen aan het opgewerkte materiaal. Andere specifieke keuzes voor SGR ten behoeve van duurzame kernenergie zijn energieomzetting met hoog rendement met een direct gekoppelde gasturbine, en verbeterde veiligheid door toepassing van helium als koelmiddel, een chemisch inert koelmiddel dat geen faseveranderingen kan ondergaan in de kern.

Omdat de bedrijfstemperaturen en -omstandigheden in de beoogde SGR lijken op die in Hoge-Temperatuur Reactoren (HTR), was het eerste SGR onderzoek voor dit proefschrift gericht op een uitbreiding van HTR spijtstoftechnologie, d.w.z. brandtofelementen met partikelspijtstof. Een dergelijk splijtstofconcept bleek niet aan de eisen van Generation IV reactoren te voldoen: het kweken bleek moeizaam, de temperaturegradienten in de spijtstofelementen zijn zeer hoog, en de spijtstofeconomie bleek tegen te vallen. Daarom worden twee verbeterde spijtstofelementen voorgesteld voor SGR, waarvan één een herontwerp van de klassieke TRISO HTR spijtstof betreft, en het ander een innovatief 'holle kogel' ontwerp. Beide spijtstofontwerpen worden toegepast met directe koeling. Om de marges voor de neutronica te vergroten en om voldoende spijtstof te kweken, heeft de voorgestelde reactor een vermogen van 2400 MWth bij een vermogensdichtheid van 50 MW/m^3. Met deze reactor is het mogelijk een gesloten spijtstofcyclus te verkrijgen. De actinidenfractie in de spijtstof blijft beperkt tot ongeveer 1%. Lange bestralingsintervallen zijn mogelijk met een laag reactiviteitsverlies, waardoor de benodigde overreactiviteit van de verse splijtstof beperkt kan blijven en er minder regelbehoefte tijdens het bedrijf is.

In de gesloten spijtstofcyclus is het noodzakelijk te kunnen voorspellen of een gegeven spijtstofsamenstelling voor cyclus i na bestralen, afkoelen en opwerken, een goede spijtstof voor cyclus $i + 1$ oplevert . Een theoretisch raamwerk wordt gepresenteerd in dit proefschrift om dit 'kweekrendement' te berekenen. Als het kweekrendement gelijk aan nul gemaakt kan worden, kan de spijtstofcyclus gesloten worden. Om het kweekrendement van de spijtstof te berekenen is het nodig te weten hoe ieder isotoop bijdraagt aan de totale reactiviteit van de reactor. Het wordt voorgesteld deze 'reactiviteitsgewichten' te berekenen met eerste-orde *eigenwaarde*perturbatietheorie. De resulterende uitdrukking kan vereenvoudigd worden tot een algemeen gebruikte definitie van reactiviteitsgewicht. Alle stappen in de spijtstofcyclus, d.w.z. bestralen, afkoelen en opwerken, moeten meegenomen worden in de berekening van het kweekrendement. Eerste-orde *nucliden*perturbatietheorie is een effectieve methode om de effecten van initiële variaties op het kweekrendement te berekenen. De theorie wordt toegepast op de gesloten spijtstofcyclus van een 600 MWth SGR. Het resultaat is dat de spijtstofcyclus gesloten kan worden als het opwerkrendement voor de gebruikte spijtstof hoog genoeg is (> 99%). Berekeningen zijn uitgevoerd met extra actiniden in de spijtstof om het transmutatiepotentieel van de SGR te schatten. Extra actiniden in de spijtstof verbeteren het kweekrendement en verkleinen het reactiviteitsverlies tijdens bedrijf.

De vermogensdichtheid in de kern van een SGR is hoog in vergelijking met andere gasgekoelde kernreactoren. Zoals elke kernreactor produceert de SGR na afschakeling nawarmte, die onder alle omstandigheden moet worden afgevoerd. Het ontwerp van de SGR beoogt Nawarmtetransport (NWT) met natuurlijke convectie waar mogelijk, met een grote koelmiddelfractie in de kern om stroming te vergemakkelijken. Door de combinatie van hoge vermogensdichtheid en lage thermische inertie in de kern zijn transiënten met hoge temperaturen mogelijk in de SGR. Om de vermogensproductie onder alle omstandigheden onder controle te houden zijn passieve elementen onderzocht om het reactorvermogen te beperken wanneer alle andere (actieve) controle-elementen falen. De elementen die worden voorgesteld gebruiken passief ingebracht vloeibaar ^{6}Li als absorber. De elementen worden door smeltzekeringen geactiveerd, en zijn voldoende klein om geïntegreerd te worden in de standaard controle-elementen. Het wordt aangetoond dat de passieve elementen het reactorvermogen op adequeate wijze kunnen beperken onder ongevalssituaties. Tevens wordt aangetoond dat NWT met natuurlijke convectie mogelijk is onder nominale druk van het primair systeem.

Delft, Oktober 2006,

W.F.G. van Rooijen

Acknowledgments

Then, finally, the moment you've all been waiting for. This part is, in a sense, the most important part of the thesis, as it attracts the largest number of readers. Dear reader, consider reading the rest of this thesis as well. It is seriously interesting, well-written, in short: a genuine master-piece! So now I've done my self-promotion, let's see who else was involved with this thesis. First and foremost, I want to thank Professors Van Dam and Van der Hagen for giving me the opportunity to work on the GCFR. It has been an interesting four years, even though I think we all agree that serious doubt remains whether a Generation IV GCFR will ever be constructed. Next, I would like to especially thank my 'daily supervisor' Dr. Jan Leen Kloosterman. With Jan Leen's (large amount of) help, advice and (especially) computer programs I wouldn't have come far. Jan Leen, you have taught me a lot over the past four years, and I'm grateful for that. Your insights and guidance have helped me many times when I had the feeling I was running around in circles or stumbling in the dark, both. I hope we can work together in the future.

I am also deeply indebted to Dr. Danny Lathouwers, although officially he had nothing to do with my thesis research. Danny, thank you for everything: assistance developing the LOWFAT code, advice in general (about almost anything) and much more. Without your experience with numerical tools I don't think our adjoint-capable transmutation code would exist. Your straight to the point advices, together with Jan Leen's theoretical guidance have given me the confidence over the past years that there is no problem in reactor physics that cannot be tackled, given enough time (note this last clause, haha). Apart from that, I want to thank you for joining me every now and again at the Sportcentrum to go 'spinning'.

Talking about spinning, I want to thank one person in particular, and that is Jelle Schut, also known as 'Snelle Jelle' or 'Jelle à Grande Vitesse'. My research had nothing to do with Jelle's work, and apart from that Jelle retired when I was in my second year. But when I started my (much needed) transformation from Obélix to Astérix, Jelle took me to the sports center to go spinning. He told me to keep it up, and to keep going even though I wasn't any good at it. In the end, Jelle's help and support gave me the motivation to keep on going, and in the end, everything worked out pretty nice. I would like to also thank Jan van Laren, the spinning instructor.

Being a PhD student can be a lonely life, but not so in the group Physics of Nuclear reactors. Respect goes out to Brian Boer ('Even een stroopschijf in m'n giegel duwen'), the only roommate who lasted longer than one year in my company. Do not be fooled, dear reader, for this is a truly remarkable achievement. Brian is the only one of four roommates (!) over the years who didn't leave the group prematurely. Godart van Gendt, my only master

student, made me realize that the task of supervising is in fact quite a demanding job. I want to mention Piet de Leege for explaining me how to make AMPX-libraries and the inner workings of NJOY and NSLINK, and for many discussions on nuclear data and what you can (or cannot) do with it.
Special acknowledgments go out to Ine Olsthoorn, for being a gentle colleague and a good friend. I will destroy that elephant some day! Although the group was somewhat stirred and rearranged during the four years I was working there, some constant factors remain: Dick de Haas, the socio-historical-cultural conscience of the group and resident expert on seventies pop music. Also thanks to Camiel Kaaijk. Thanks to him a part of Dutch vocabulary will never be the same (Hilti, roodband, knikkerbak). I should not forget to mention Sieuwert-sieuwert (slightly insane but a good friend), Stavros (slightly more insane), Christian, Carlos and August. Thanks for everything: laughs, good conversation during coffee breaks and lunches, and so much more.
I also want to thank the group who hosted me during my internship at CEA Cadarache (ehhh-mmm let me think: CEA/DEN/CAD/DER/SESI/LCSI, probably the longest acronym I have ever seen for a lab). In no particular order, thanks to: Jean-Claude Garnier, Patrick Dumaz, Jean-Yves Malo (I hope I didn't disappoint him), Christophe Bassi (taught me CATHARE, thanks for that), Frank Morin (hmmm. Wine, anybody?), and others. I had a very good time in France, and I learnt a lot about practical reactor engineering. I am sure we will meet again in the future, because after all, it's a small, small world.
I would like to acknowledge the support of the European Commission. My salary was paid (partly) by the GCFR STREP, carried out under Contract Number 012773 (FI6O) within the EURATOM 6th Framework Programme (`http://www.cordis.lu/fp6/`), effective from March 1st 2005 to February 28th 2009. More information about the GCFR STREP is available from the project website (`http://www.gcfr.org`). I would like to thank the STREP partners for valuable discussions and contributions to my research.

One person deserves a special mention, and that is Paul Kruis, who gradually evolved from my girlfriends' flatmate to a dear friend. And now that we are at it, I'd like to say hello to my parents.

A very special thanks to my wife Edith, for all her support, care and attention. Even when we are miles apart, you will always be with me in heart and thoughts.

W.F.G. van Rooijen
M. van Hennebergweg 1[B]
2552 BA Den Haag
The Netherlands

Born: June 11th, 1977
in Haarlem, The Netherlands
Nationality: Dutch

Curriculum Vitae

Education

2002 - 2006 • Ph.D. research Gas Cooled Fast Reactor, fuel cycle and safety. Delft University of Technology, The Netherlands

2001 - 2002 • Post-graduate study 'Japanese Business and Culture', Leiden University (The Netherlands), Japan-Netherlands Institute and AIST (Japan)

1995 - 2001 • Study: Applied Physics at Delft University of Technology, The Netherlands. Title awarded: M.Sc.

1989 - 1995 • Preparatory Scientific Education (VWO) at Stedelijk Gymnasium Haarlem, The Netherlands

Extra-curricular

2002 • Radiation Protection and Hygiene Expert level III

2002 • JETRO Business Japanese Language Proficiency level III

Professional Experience

2007 - ... • Assistant professor at George W. Woodruff School of Mechanical Engineering Nuclear and Radiological Engineering / Medical Physics Program, Georgia Institute of Technology, USA

2002 - 2006 • Employee of Delft University of Technology as Ph.D. student, Dept. Physics of Nuclear Reactors, specialism: Gas Cooled Fast Reactor

2004 - 2006 • Supervisor for practical work for Radiation Hygene Course at Utrecht University, The Netherlands (in scope of Ph.D. contract)

2001 • Programmer, Laboratory for Acoustic Imaging and Sound Control, Delft University of Technology

Research related education & skills

Educational courses followed

2005 • Course Reactor Physics: Kinetics of Nuclear Reactors at Delft University through ENEN (European Nuclear Education Network)

2004 • Course Reactor Physics: Dynamics of Nuclear Reactors at Delft University

2003 • Expert course in basic reactor physics with practical work in Mol, Belgium (ENEN)

2003 • Course Reactor Physics Special Topics at Delft University

Research related

- Thorough knowledge of fuel cycle physics, reprocessing strategies and related fields. Thorough knowledge of neutronics and neutronics calculations, especially for fast, gas cooled systems.

Nuclear Code Packages

- Thorough knowledge of SCALE system and its modules for neutronics analyses; knowledge of MCNP, NJOY, and CATHARE2.

Programming

- Knowledge of Perl, Fortran, LaTeX; familiar with Matlab, SciLab, gMake etc.

Miscellaneous

Internships

2004 • Internship at CEA, Cadarache, France. Design of the primary helium - water heat exchanger for ETDR, and transient analysis of ETDR's primary circuit with heat exchanger using thermo-hydraulics code CATHARE2 (4 months)

2002 • Internship at AIST Chuo, Tsukuba, Japan. Optimization of automatic speech recognition for a mobile office robot using a Support Vector Machine. Implementation of SVM in a simulation (Matlab) and actual DSPs (7 months)

2000 • Internship at Volkswagen AG, Wolfsburg, Germany. Optimization of piston geometry for a direct injection turbo-diesel engine (4 months)

Languages

- Dutch (native), English (fluent), Japanese (quite good), French (good), German (good)